Dynamic Stability of the Bedding Rock Slope with Deteriorated Hydro-fluctuation Belt

Xinrong Liu · Bin Xu

Dynamic Stability of the Bedding Rock Slope with Deteriorated Hydro-fluctuation Belt

Springer

Xinrong Liu
Chongqing University
Chongqing, China

Bin Xu
Chongqing Jiaotong University
Chongqing, China

ISBN 978-981-95-6310-4 ISBN 978-981-95-6311-1 (eBook)
https://doi.org/10.1007/978-981-95-6311-1

This work was supported by National Natural Science Foundation of China.

This Springer imprint is published by the registered company Springer Nature Singapore Pte Ltd.
The registered company address is: 152 Beach Road, #21-01/04 Gateway East, Singapore 189721, Singapore

Preface

The Three Gorges Project (TGP) has enormous comprehensive benefits such as flood control, power generation and navigation. However, after the completion of the project, the disasters of the bedding rock slopes (BRSs), which triggered by the frequent occurrence of the reservoir induced earthquakes (REs) and the increasing deterioration of rock mass in the hydro-fluctuation belt (HB) in the Three Gorges Reservoir Area (TGRA), have posed a significant threat to the safety of people's lives and property. In this regard, the Central Committee of the Communist Party of China and the State Council attach great importance to the prevention and control of such the disasters of the BRSs and point out that the construction of the Yangtze River Economic Belt urgently requires the construction of a safe ecological reservoir channel, providing strong guarantees for a series of major national strategic needs. At present, the investigation on key scientific issues such as the fatigue damage mechanism and strength degradation model of the rock joints of the BRS under complex conditions, the dynamic response rules and instability evolution mechanism of the BRS, the factors influencing the dynamic stability of the BRS and their sensitivity levels are not yet comprehensive enough. A series of related basic study needs to be carried out in a deep and systematic manner to fill the research gap in this field and promote its further application and development.

The main research content of this book is organized as follows, i.e., the analysis of the in-site investigation and research status in Chap. 1, the development of the cyclic shear test equipment and the study of macro-meso fatigue damage mechanism of the rock joints in Chap. 2, the construction of a strength degradation model for the rock joints and its numerical solution method in Chap. 3, the development of the shaking table model test equipment and the study of dynamic response rules and instability evolution mechanism of the BRS in Chap. 4, the study of on the factors influencing the dynamic stability of the BRS and their sensitivity levels in Chap. 5. Specifically, this book takes the common BRSs in engineering practice of the TGRA as the research object. The self-developed new test equipments have been adopted in this book, and a combination approach of in-site investigation, indoor experiments, theoretical analysis and numerical development has been adopted. Based on this, the characteristics of the REs, deterioration of rock mass in the HB and the disasters of the BRSs have been clarified, the macro-meso fatigue damage mechanism of the rock joints subjected to pre-peak cyclic shear loading has been explored, the dynamic stability solution model for the BRS considering the strength degradation effect of the rock joints under vibration has been established, the dynamic response rules and instability evolution mechanism of the BRS considering deterioration of rock mass in the HB have been revealed, and the influence rules and sensitivity levels of multiple factors on the dynamic stability of the BRS have been analyzed after introducing the strength degradation model of the rock joint. This book adopts a research system of in-site investigation, equipment development, experimental exploration, theoretical construction and numerical development, with detailed information, rich content and evidence-based arguments. It is an effective combination of scientific theory and engineering practice. The relevant research approaches and achievements can lay a theoretical foundation and provide scientific basis for solving the key scientific problems behind the prevention and control technologies of the disasters of the BRSs and will have great economic, social, environmental benefits and broad application prospects. We sincerely hope the publication of this book will be beneficial for the healthy and sustainable development and application of the disaster prevention and control system for the reservoir BRSs

This book integrates the research achievements of multiple scientific and technological workers in recent years. Specifically, with the support of the National Natural Science Foundation "Research on static (dynamic) force instability mechanism and stability control of slope under the condition of rock mass deterioration of the hydro-fluctuation belt in the reservoir area (Grant No. 41972266)", the author has trained one doctoral student and five master's students successively. Bin Xu (Ph.D. thesis "Study on dynamic damage and instability mechanism of bedding reservoir bank slope under the influence of rock mass deterioration in the hydro-fluctuation belt"), Rui Jing (Master's thesis "The deformation and failure modes of the fluctuation belt of reservoir bank slopes and stability analysis and evaluation of reserved bedding rock slopes"), Guangyi Lin (Master's thesis "Study on the cumulative damage mechanism and stability factors of soft and hard interbedded slope under cyclic shear"), Junhui Huang (Master's thesis "Study on dynamic response characteristics and stability of bedding rock slope in water level fluctuating zone of reservoir area"), Xi Zeng (Master's thesis "Study on dynamic response characteristics and stability of typical dangerous rock mass along the Three Gorges Reservoir"), Xinlin Liu (Master's thesis "Study on the stability of limestone rock shore slope under rock degradation in hydro-fluctuation belt"). On the basis of summarizing and sorting out the above research results, this book has been formed. We are very grateful for their contributions to this book. The research results of this book have also been supported by the National Natural Science Foundation of China (Grant No. 52509138), the China Postdoctoral Science Foundation (Grant No. 2023M730432), the Special Funding for Chongqing Postdoctoral Research Project (Grant No. 2022CQBSHTB1010), and the Chongqing Postdoctoral Science Foundation (Grant No. CSTB2023NSCQ-BHX0223). We also cite some references from peer experts and would like to express our sincerest gratitude.

Due to the authors' ability, there are inevitably omissions and shortcomings in the book. Readers are kindly requested to criticize and correct them.

Chongqing, China
October 2025

Xinrong Liu
Bin Xu

Competing Interests The authors have no competing interests to declare that are relevant to the content of this manuscript.

Abbreviations and Symbols

The following list of abbreviations and symbols is provided for ease of reference for those abbreviations and symbols that are most frequently used in this book. It is therefore not exhaustive. Nonetheless, all abbreviations and symbols used in this book are defined at their first mention in the main text.

Abbreviations

$ASCM_{min}$	Minimum average safety coefficient method
BCQs	Basic control quantities
BP_{DA}	Dip angle of bedding plane
BP_{T}	Thickness of bedding plane
BRSs	Bedding rock slopes
C_{D}	Cumulative displacement
C_{DG}	Cumulative damage degree
CNL	Constant normal loading
DCD	Dynamic cumulative damage
DL_{A}	Dynamic load amplitude
DL_{F}	Dynamic load frequency
D_{R}	Damping ratio
DR_{S}	Degradation ratio of shear strength
D_{SC}	Dynamic stability coefficient
DSC_{max} and DSC_{min}	Maximum and minimum dynamic stability coefficient, respectively
DVM	Dynamic vector method
FPM	Force polygon method
GI_{S}	Gently inclined slope
HB	Hydro-fluctuation belt
HB_{DR}	Degradation range of the HB
HB_{DS}	Degradation shape of the HB

HB_{HW}	Height of water-level of the HB
HD	Horizontal distance
ISF	Instability failure
LET	Limit equilibrium theory
MI_S	Moderately inclined slope
MLT	Mass—length—time
N_C	Number of cycles
N_F	Natural frequency
NW	Natural wave
OAM	Orthogonal analysis method
PFC^{2D}	Particle flow code 2D
PGA	Peak-ground acceleration
P_W	Pore-water pressure
REs	Reservoir induced earthquakes
R_P	Rock pressure
S_A	Slope angle
S_H	Slope height
SIF	Seismic inertial force
SI_S	Steeply inclined slope
S_{SC}	Static stability coefficient
SVDC	Strength vibration degradation coefficient
$SVDC_{min}$	Minimum strength vibration degradation coefficient
SW	Sine wave
TGD	Three Gorges Dam
TGP	Three Gorges Project
TGRA	Three Gorges Reservoir Area
UDEC	Universal distinct element code
VD	Vertical distance
WN	White noise
W_N	White noise serial number

Symbols

τ_p	Average value of the peak shear strength
$\tau_{n\text{-}s\text{-}d}$, $\tau_{e\text{-}s\text{-}d}$	Stress drop
$\tau(t)$	Shear strength of the rock joints at any moment t of seismic duration
$\sigma(t)$	Normal stress acting on the rock joints at any moment t of seismic duration
$c(t)$	Cohesion of the rock joints at any moment t of seismic duration
$\varphi(t)$	Internal friction angle of the rock joints at any moment t of seismic duration

$D_{sc}(t)$	SVDC of the rock joints at any moment t of seismic duration
τ_0	Initial shear strength of the rock joints
c_0	Initial cohesion of the rock joints
φ_0	Initial internal friction angle of the rock joints
α	Asperity angle of the rock joints
$\xi(t)$ and $\lambda(t)$	Influence factor of vibration wear and relative motion velocity of the rock joints at any moment t of seismic duration, respectively
$\omega(t)$	Convergence value of the influence factor of vibration wear of the rock joints at any moment t of seismic duration
m, n and k	Undetermined coefficient
$N(t)$	Number of cycles of the rock joints at any moment t of seismic duration
P_0 and Q_0	Convergence value of the influence factor of vibration wear and relative motion velocity of the rock joints at any moment t of seismic duration, respectively
$A(t)$	Cyclic shear amplitude of the rock joints at any moment t of seismic duration
$V(t)$	Relative motion velocity between rock blocks at any moment t of seismic duration
$\overrightarrow{R_f}(t)$	Reaction force vector in the normal direction of the sliding surface at any moment t of seismic duration
$\overrightarrow{R_{ef}}(t)$	Combined external force vector of the slip mass at any moment t of seismic duration
$\overrightarrow{m}$	Normal unit vector outside the sliding surface
$\overrightarrow{S_f}(t)$	Total sliding force vector of the slip mass at any moment t of seismic duration
$\overrightarrow{n}$	Tangential unit vector of the sliding surface
$F_s(t)$	Instantaneous D_{SC} of slope at any moment t of seismic duration
A	Contact area between the slip mass and sliding surface
$F_v(t)$ and $F_h(t)$	Vertical and horizontal SIF of the slip mass at any moment t of seismic duration, respectively
G	Self-gravity of the slip mass
h	Height of the slip mass
θ	Inclination angle of the sliding surface
$\overrightarrow{F_{udec}}(t)$	Combined external force vector of the slip mass at any moment t of seismic duration, respectively
n_j	Number of grid nodes of the slip mass

Symbol	Description
$\overrightarrow{F^{g}_{udec}}(t)$	Nominal gravity vector of grid nodes of the slip mass at any moment t of seismic duration, respectively
$\overrightarrow{F^{l}_{udec}}(t)$	External load vector of the slip mass at any moment t of seismic duration
$\overrightarrow{F^{z}_{udec}}(t)$	Internal force vector of the grid element near node of the slip mass at any moment t of seismic duration
$\overrightarrow{F^{c}_{udec}}(t)$	Sub contact force vector only existing between rock blocks of the slip mass at any moment t of seismic duration
$\overrightarrow{a}(t)$	Integral acceleration vector of the slip mass at any moment t of seismic duration
m_r	Mass of the slip mass
$\overrightarrow{F_{hv}}(t)$	Instantaneous SIF vector of slope at any moment t of seismic duration
γ_{eq}	Equivalent and constant hypothetical shear strain
ϑ	Conversion coefficient
$\gamma_{\max}$	Maximum value of the element shear strain in the seismic duration
F_{dei}	Evaluation index of the dynamic stability of slope
F_{s0}	S_{SC} of slope
$F_{d\min}$	Minimum stability coefficient in the time history curve
a	Acceleration of the loading wave
λ' and λ	Amplitude of the loading wave
f'	Frequency of the loading wave
t'	Duration of the loading wave
$R_x(t)$ and $R'_x(t)$	Reaction force of block i against $i-1$ and block $i+1$ against i at any moment t of seismic duration, respectively
$F_{h_{i-1}}(t)$, $F_{h_i}(t)$ and $F_{h_{i+1}}(t)$	Horizontal SIF of block $i-1$, i and $i+1$ at any moment t of seismic duration, respectively
$R_{i-1}(t)$, $R_i(t)$ and $R_{i+1}(t)$	Reaction force of the rock mass at the lower part of block $i-1$, i and $i+1$ against its sliding surface at any moment t of seismic duration, respectively
$F_{v_{i+1}}(t)$	Vertical SIF of block $i+1$ at any moment t of seismic duration
l_{i-1}, l_i and l_{i+1}	Length (area per unit width) of the sliding surface of block $i-1$, i and $i+1$, respectively
l'_x	Adjacent interface length (area per unit width) of block $i+1$ and i
θ_{i-1}, θ_i and θ_{i+1}	Inclination angle of the sliding surface of block $i-1$, i and $i+1$, respectively

$N_{i-1}(t)$, $N_i(t)$ and $N_{i+1}(t)$	Normal force outside the sliding surface of block $i-1$, i and $i+1$ at any moment t of seismic duration, respectively
$F_{v_{i-1}}(t)$ and $F_{v_i}(t)$	Vertical SIF of block $i-1$ and i at any moment t of seismic duration, respectively
G_{i-1}, G_i and G_{i+1}	Self-gravity of block $i-1$, i and $i+1$, respectively
$F_{si}(t)$	Assumed instantaneous D_{SC} that constantly updated by the numerical program
ρ, E, l, c, φ, σ, ε, μ, t f and g	Physical attribute of density, elastic modulus, geometric size, cohesion, internal friction angle, normal stress, axial strain, Poisson's ratio, time, frequency and gravitational acceleration, respectively
C_ρ, C_E, C_l, C_c, C_φ, C_σ, C_ε, C_μ, C_u, C_t, C_f, C_v, C_g and C_ζ	Similarity coefficient of density, elastic modulus, geometric size, cohesion, internal friction angle, normal stress, axial strain, Poisson's ratio, displacement, time, frequency, velocity, acceleration and damping ratio, respectively
σ_c	Compressive strength
τ	Shear stress
F	Indicator in the F-distribution
α_x	Significance level

Contents

Chapter 1
In-situ Investigation and Analysis

1.1 Introduction

China is one of the countries with the most severe geological disasters and the largest population threatened by them in the world. The high risk areas of geological disasters in China are mainly distributed in the southwest high mountain areas, southeast low mountain and hilly areas, the loess plateau, and Liaodong low mountain and hilly areas, with landslide disasters being the most significant (Yang et al., 2020; Guo & Yin, 2021). Specifically, as the largest water conservancy hub project in China and even in the world, the TGP has formed a narrow and long river shaped reservoir from the beginning to the end, including counties of Zigui, Badong, Wushan, Fengjie, Yunyang, etc., with a main and tributary length of 630 km and 5000 km, respectively, and an area of 1084 km^2. Moreover, due to the large population surrounding the TGRA, the vegetation coverage on both sides of the strait has decreased to 5%, so its geological environment capacity is extremely limited (Fourniadis et al., 2007; Ren, 2010; Cheng, 2018; Tang et al., 2019). According to incomplete statistics, there are over 20000 geological disasters in the reservoir area, mainly distributed on the slopes on both sides of the reservoir area. As a result, the bank slope with length of 103 km is in a state of poor stability throughout the year. Once the conditions are ripe, disasters may occur at any time, with over 20 million people affected by the disasters (Zhu, 2010; Song, 2019; Ye et al., 2021). In particular, the Central Committee of the Communist Party of China and the State Council have always attached great importance to the prevention and control of geological disasters on the bank slopes of the reservoir area, and pointed out that this work is a national long-term plan. The golden waterway of the Yangtze River plays a highly supportive role in China's social and economic development, and the construction of the Yangtze River Economic Belt urgently requires the construction of a safe and ecological reservoir waterway. It provides a strong guarantee for national major strategic needs such as "lucid waters and lush mountains are invaluable assets", "to step up conservation of the Yangtze River and stop its over development" and "the Belt and Road".

X. Liu and B. Xu, *Dynamic Stability of the Bedding Rock Slope with Deteriorated Hydro-fluctuation Belt*, https://doi.org/10.1007/978-981-95-6311-1_1

Since the impoundment of the TGRA in 2003, the original geological environment in the area affected by the narrow and elongated reservoir has been disturbed or damaged to a certain extent, resulting in frequent REs and increasing deterioration of rock mass in the HB. Furthermore, the geological disasters such as sliding, collapsing, toppling, falling, subsiding and seismic cracking of the bank slopes, which are of various types, large-scale, high frequency, widely distributed and severely harmful (especially common in the BRSs), often occur (Gu, 2018; Huang et al., 2020; Ye et al., 2021). In view of this, the scientific research team has continued to carry out a large number of in-site investigation work in the reservoir area in recent years. Based on a comprehensive analysis of existing studies and in-site investigation data, it can be concluded that the REs that have occurred in the reservoir area since the first water storage are mainly in the form of small earthquake releases. Meanwhile, the REs have a series of characteristics such as shallow source, short duration, high seismic frequency, high ground peak acceleration, strong epicenter intensity, fast intensity attenuation and small range of extreme earthquake areas, and the number of earthquakes is increasing year by year (Zhao, 2009; Chen et al., 2010; Liu, 2017; Wang, 2017; Liu et al., 2020a). In addition, the HB widely distributed in all counties in the reservoir area of Hubei Province and Chongqing City is a typical water rock interaction area (with an area of 400 km^2). The rock mass in this area is in a long-term dry wet cycle state under periodic fluctuations of reservoir water level of 145 ~ 175 m, and its rock mass deterioration degree accumulates and develops with the increasing number of water level fluctuations (dry wet cycles) (Deng, 2010; Fu, 2010; Fang et al., 2021). In particular, this condition causes irreversible deterioration effects such as soaking (air drying), erosion (stripping) and dissolution (submergence) on the rock mass of the BRSs commonly distributed on both sides of canyon river type reservoirs, resulting in the gradual fragmentation of the rock mass in the HB and the emergence of a large number of serious damage phenomena such as cracks, hollows, dissolution channels and cavities. In addition to the influence of frequent REs, the stability of the BRSs with severe rock mass deterioration in the HB continues to decrease until instability and failure occur (Huang et al., 2019; Liu et al., 2020b; Xu et al., 2022a). From this, it can be seen that the frequent occurrence of the REs in the reservoir area and the worsening of rock mass deterioration in the HB pose a significant threat to the stability of the BRSs. Moreover, the series of geological disasters caused by the combined influence of the two have posed serious safety hazards to the lives and property of residents, transportation and shipping, as well as power station facilities in the reservoir area. Therefore, high attention should be paid to this issue.

To sum up, with the aim of actively responding to major national strategic needs and committed to solving the key scientific issues behind the prevention and control technology of geological disasters on the bank slopes of the TGRA, and effectively considering the new situations, problems and challenges faced by the TGRA, it is necessary to systematically analyze the REs, rock mass deterioration in the HB and landslide disasters on the BRSs in the reservoir area based on in-site investigation and analysis. To lay a research foundation for further exploring the dynamic damage mechanism and stability of the BRSs under the deterioration of rock mass in the HB.

The relevant achievements can lay a theoretical foundation and provide scientific basis for the following situations, such as analyzing and evaluating the long-term dynamic stability of the BRSs in complex geological environments, enhancing the comprehensive capabilities of geological disaster prevention, reduction and relief in the reservoir channel, ensuring the safety of the lives and property of residents in the reservoir area, and guaranteeing the long-term safe operation of the Three Gorges Dam (TGD). It will have great economic, social and environmental benefits, as well as great application prospects.

1.2 REs in the TGRA

After the completion of the TGD and the impoundment of the reservoir, various geological catastrophes often occur, which can threaten the safety of shipping and surrounding residents in the TGRA. Specifically, the analysis of data from the in-situ investigation and available studies (Yao et al., 2017a; Huang et al., 2018; Kang et al., 2019; Xu et al., 2021; Xu et al., 2022a, b, c) showed that the number of the RIEs in the TGRA had increased yearly since the first impoundment in 2003. For example, there were more than 500 seismic waves in 2003 (an increase by a factor of 3 ~ 13 times over the previous seven years), there were more than 1000 seismic waves in 2004 (approximately twice the number in the previous year), and there were more than 5000 seismic waves in the first section in the TGRA in 2009 and 2010 (an augmentation of approximately 10 folds compared to that in 2003). The distribution situation of the RIEs in the TGRA (Zigui County → Badong County) from 1971 to 2015 is shown in Fig. 1.1. In particular, the frequency of occurrence of the RIEs has remained at a high level since impoundment operation at 175 m elevation in 2008, and several RIEs with large magnitude have occurred in the short term, e.g., grade 4.1 (G4.1, similarly hereinafter) in 2008, G5.1 in 2013, G4.2 and G4.5 in 2014, and G4.1 and G4.3 in 2017.

1.3 Deterioration of Rock Mass in the HB

The water-level in the TGRA fluctuated periodically from 145 to 175 m (as shown in Fig. 1.2) year-round, causing irreversible damage and deterioration of rock mass in the HB of the BRSs (most widely distributed in the TGRA), and these damage and deterioration effects will gradually increase with time. Specifically, for the deterioration modes of rock mass in the HB and the geological structure characteristics of the bank slopes in the key research area (the Wushan section in the TGRA), the analysis of the data from the in-situ investigation and previous studies (Deng et al., 2013; Gu et al., 2017; Wan et al., 2017; Tang et al., 2019; Wang et al., 2019; Huang et al., 2021; Xu et al., 2022a, c) showed that the typical deterioration modes of rock mass in the HB of this section could be summarized as belonging to four types, i.e.,

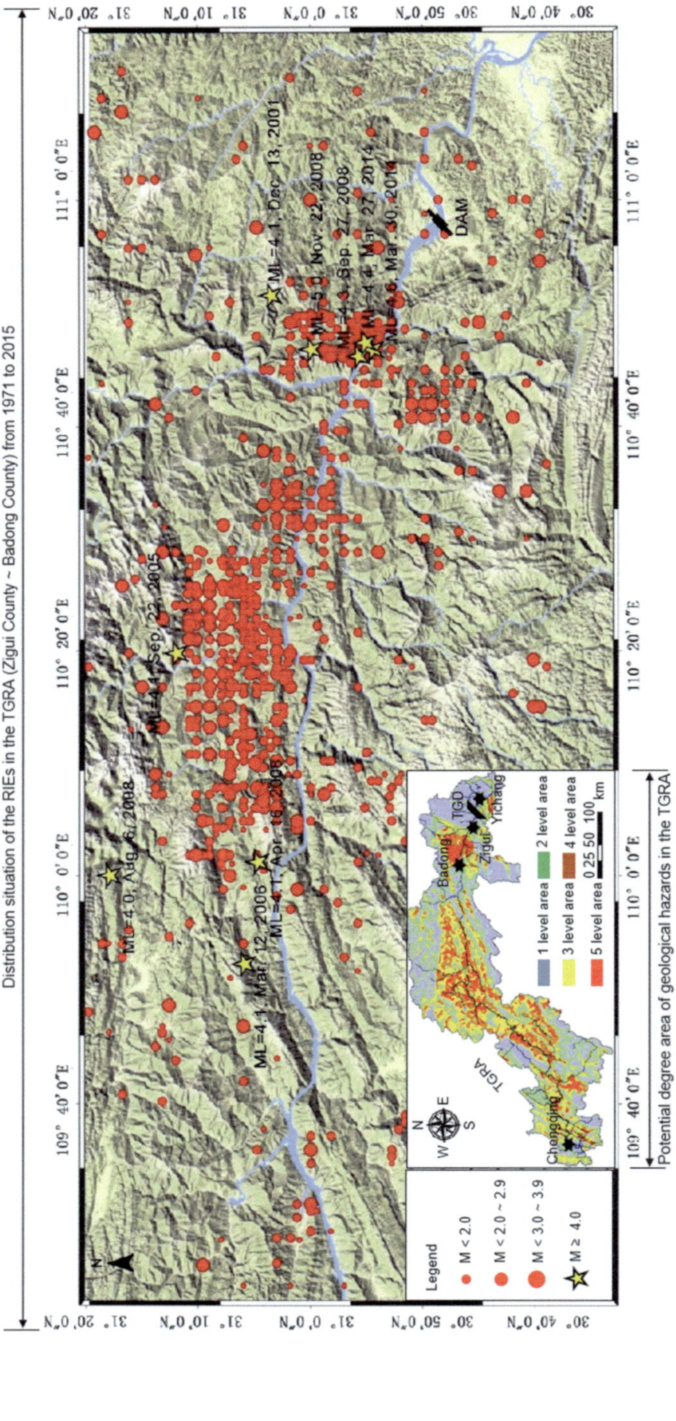

Fig. 1.1 Distribution situation of the RIEs in the TGRA

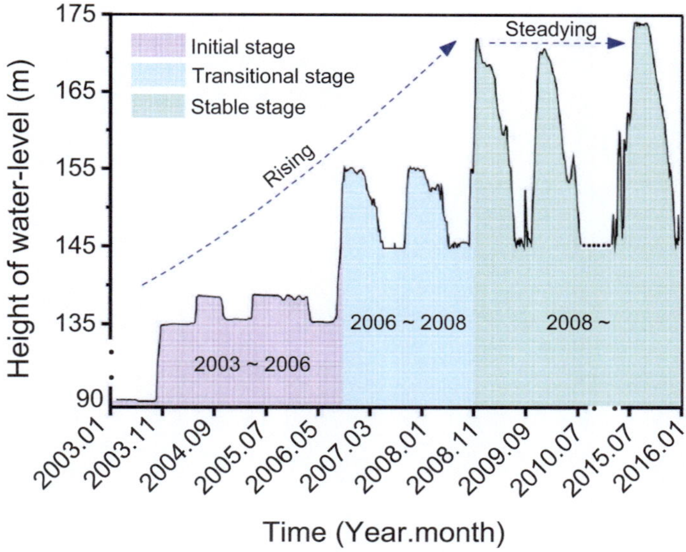

Fig. 1.2 Evolution curve of water-level fluctuation in the TGRA

erosion (denudation) type (common in the rock or rock-soil bank slopes, as shown in Mode A in Fig. 1.3), corrosion (suffosion) type (common in the rock bank slopes, as shown in Mode B in Fig. 1.3), avalanche (collapse) types (common in the rock or soil bank slopes, as shown in Mode C in Fig. 1.3) and slip (flow) type (common in the soil bank slopes, as shown in Mode D in Fig. 1.3).

For Mode A, the rock-soil mass fissures in the HB are extremely developed and the bank slope gradually moves backwards; for Mode B, the local rock cavity develops following minor suffosion and corrosion of rock mass in the HB; for Mode C, the rock-soil mass in the HB is in a free state after partial avalanche and collapse; and for Mode D, the front edge of the soil mass in the HB is partially subject to overall slip or flow. For Modes A ~ D, the original shape of the bank slope is replaced by a new slope shape after the rock mass in the HB gradually deteriorates. It should be emphasized that the macro-scale deterioration form of rock mass in the HB of this section is dominated by erosion (denudation) type, i.e., the distribution of the broken rock blocks in the HB is approximately reticulated and interlaced (the fissures were extremely developed), and its main failure characteristics are shown as cutting, loosening, blocking and sliding of rock mass (the bank slope line moved slowly backwards).

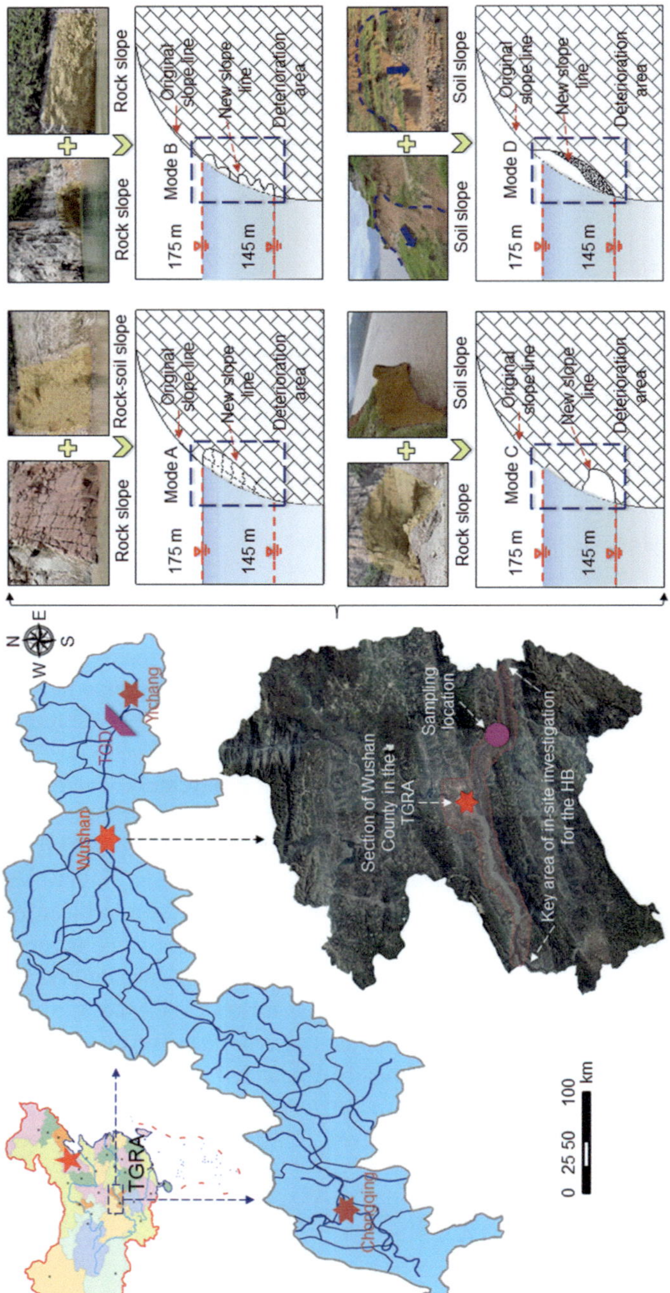

Fig. 1.3 Typical deterioration modes of rock mass in the HB of the TGRA

1.4 Landslide Disasters of the BRSs

China is one of the countries with the most severe geological disasters and the most threatened population in the world. Geological disasters are widely distributed, with a high proportion of landslide disasters and a significant impact. Especially, the geological disaster situation along the TGRA is extremely severe (as shown in Fig. 1.4 for in-site investigation pictures of various existing or potential bank slope disasters in the TGRA), with the landslide disasters of the BRSs being the most obvious. Considering the impact of rock mass deterioration in the HB, in addition to the impact of frequent RIEs, the overall stability of the bank slope with deteriorated HB continuously decreased and even destabilization disasters occurred. The BRSs in the TGRA are mainly distributed in Zigui, Badong, Wushan, Fengjie, Yunyang and other counties. The evaluation of the in-situ investigation data and previous studies (He & Wang, 2006; Jian et al., 2014; Xu et al., 2015; Wan et al., 2017; Huang et al., 2019; Lu & Huang, 2020; Yin et al., 2020; Zhang et al., 2021; Yang et al., 2022) showed that landslide disasters of various types, large-scale and severe damage occasionally occurred in this type of the BS (as shown in Fig. 1.5), and the stability of this type of the bank slope is usually significantly affected by the characteristics of its bedding planes.

Particularly, Nos. 1* ~ 15* are the locations of landslide disasters in Fig. 1.5, i.e., Nos. 1* ~ 15* represent Xintan landslide in Zigui County, Xirangpoxi landslide in Badong County, Maliuzui landslide in Banan District, Wufengshan landslide in Yunyang County, Xicheng landslide in Yunyang County, Shizibao landslide in Badong County, Baijiabao Landslide in Zigui County, Qianjiangping landslide in Zigui County, Ji'an landslide in Wanzhou District, Outang landslide in Fengjie County, Taping landslide in Wushan County, Qingshi landslide in Wushan County, Kuangjiatuo landslide in Wushan County, Shanshucao landslide in Zigui County, and Kamenziwan landslide in Zigui County, respectively. Additionally, the drilling data showed that the deterioration depth of rock mass in the HB of this section was primarily distributed within the range of 0 ~ 15 m, and the elevation of the stable groundwater level was approximately 180 m. Additionally, the lithology of the BRSs with the fluctuation range of slope angle was 30 ~ 90° in this section was dominated by limestone, and most BSs contained medium thickness and straight bedding planes where the fluctuation ranges of the thickness and dip angle were 3 ~ 8 m and 0 ~ 90°, respectively.

1.5 Summary and Conclusions

(1) The construction of the TGP has promoted the rapid development of the Yangtze River shipping economy, ensured the safe regulation of water levels along the river during the flood season, and alleviated the tense situation of power supply in the central and southern regions of China. However, the construction of the TGP

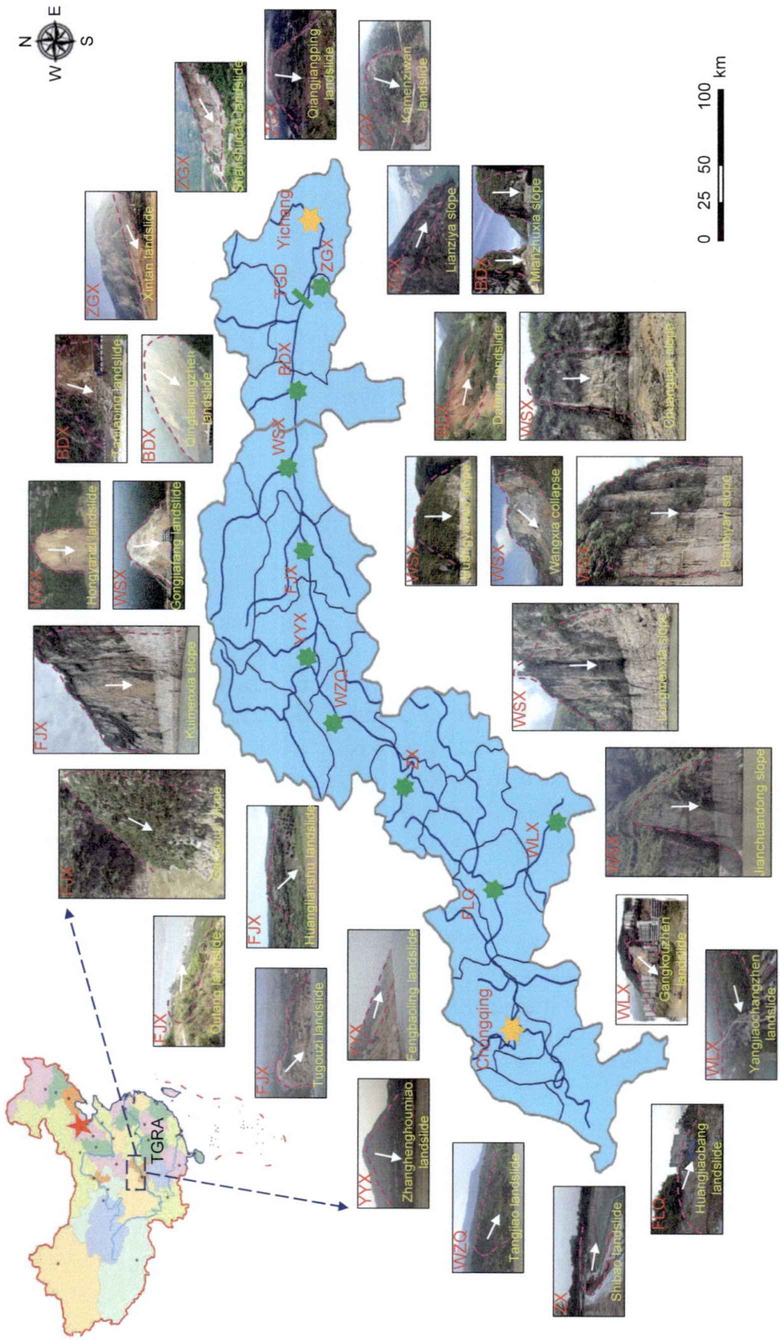

Fig. 1.4 In-site investigation pictures of various existing or potential bank slope disasters in the TGRA

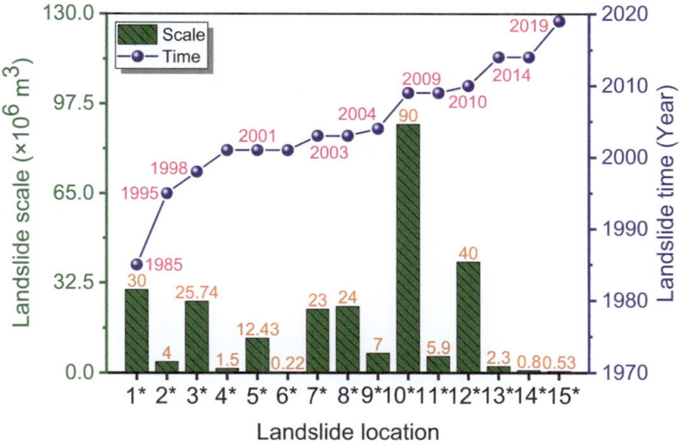

Fig. 1.5 Statistics of cases for landslide disasters of the BRSs in the TGRA

has led to an irreversible evolution of the original hydrogeological conditions in the reservoir area, mainly manifested in the severe deterioration of rock mass in the HB caused by periodic fluctuations of the reservoir water level and the frequent occurrence of the REs, which in turn have led to numerous types, large-scale, high frequency, wide distribution and serious harm of geological disasters on the bank slopes. Among them, the landslide disasters caused by the coupling effect between the deterioration of rock mass in the HB and the REs (mainly frequent small earthquakes) are the most common, posing a serious threat to the safe operation of shipping in the reservoir area, and the lives and property of urban and rural residents.

(2) The REs that have occurred in the reservoir area since the first water storage are mainly in the form of small earthquake releases. Meanwhile, the REs have a series of characteristics such as shallow source, short duration, high seismic frequency, high ground peak acceleration, strong epicenter intensity, fast intensity attenuation and small range of extreme earthquake areas, and the number of earthquakes is increasing year by year. The typical deterioration modes of rock mass in the HB of this section could be summarized as belonging to four types, i.e., erosion (denudation) type (common in the rock or rock-soil bank slopes), corrosion (suffosion) type (common in the rock bank slopes), avalanche (collapse) types (common in the rock or soil bank slopes) and slip (flow) type (common in the soil bank slopes).

(3) Numerous scholars at home and abroad have conducted many studies on the stability of the reservoir bank slopes, but most of them focus on exploring the related issues of their respective impacts on the stability of the bank slopes from the perspective of the REs or the rock mass deterioration in the HB. Especially, there are few reports on literature considering the joint influence of the two and conducting research on the stability of the BRSs. Overall, the key scientific

issues that currently exist or urgently need to be addressed in this field mainly include the following, i.e., the fatigue damage mechanism and theoretical model of the rock joints of the BRS are not clear, the dynamic response rules and instability evolution mechanism of the BRS considering deterioration of rock mass in the HB are not clear, and the influence factors of the dynamic stability of the BRS and its sensitivity order considering deterioration of rock mass in the HB are not clear. It is urgent to conduct in-depth and systematic scientific research work around the above topics.

Chapter 2
Mechanical Behaviours of Fatigue Damage of the Rock Joints Under Pre-Peak Cyclic Shear Loading

2.1 Introduction

It is well known that most natural rock masses contain structural planes such as joints, faults, fissures and weak layers, that their shear mechanical behaviours are significantly different from those of intact rock masses, and that the shear mechanical behaviours play an important role in the structural stability of rock masses (Liu et al., 2018a; Feng et al., 2018; Zhang et al., 2020). In particular, the cyclic shear load generated by earthquakes and explosions will cause fatigue damage effects, such as dislocations, fractures and slipping of the rock joints, and will lead to progressive failure of the stability of engineering rock masses (Erarslan & Williams, 2012; Liu, 2017); thus, revealing the shear mechanical behaviours of the rock joints under earthquake and explosion loading is the basic premise for the dynamic stability analysis and evaluation of engineering rock masses. In general, the shear mechanical behaviours (fatigue damage) of the rock joints under cyclic shear loading are very complex and are often affected by the rock mass type (weathering degree), joint morphology (especially important), normal stress, shear rate, shear amplitude and other factors (Barton & Choubey, 1977; Liu et al., 2018b). Therefore, it is of great theoretical and practical significance to study the fatigue damage mechanism of the rock joints under cyclic shear loading to analyse and evaluate the dynamic stability of rock masses in surface and underground engineering applications, such as slopes, foundation pits, dams and tunnels.

There are currently many studies about the influence of cyclic shear loading on the fatigue damage mechanism of the rock joints (Ge, 1979; Barton & Choubey, 1977; Lee et al., 2001; Li et al., 2003; Jafari et al., 2003; Mirzaghorbanali et al., 2014; Fathi et al., 2016; Liu, 2017; Niktabar et al., 2017; Feng et al., 2018; Liu et al., 2018b; Kou et al., 2019; Liu et al., 2019; Liu et al., 2020c). Earlier, Ge (1979) described the mechanical behaviours of the shear fatigue damage of the rock joints under repeated loading and unloading, and Jafari et al. (2003) analysed the fatigue damage mechanical behaviours of the saw-tooth rock joints (cemented with special mortar)

X. Liu and B. Xu, *Dynamic Stability of the Bedding Rock Slope with Deteriorated Hydro-fluctuation Belt*, https://doi.org/10.1007/978-981-95-6311-1_2

under different stress amplitudes and numbers of cycles. Subsequently, Mirzaghorbanali et al. (2014) and Niktabar et al. (2017) prepared the saw-tooth rock joints with high-strength gypsum and plaster of Paris, and explored the influence mechanism of different shear rates and roughnesses on the fatigue damage mechanical behaviours of the rock joints. However, most of the existing literature uses a physical model made of similar materials when studying the mechanism of cyclic shear fatigue damage of the rock joints, and the cyclic shear load is generally applied based under macroseismic conditions, i.e., each cyclic loading will exceed the peak shear strength of the rock joints and cause significant bidirectional dislocation. In particular, the existing literature shows that the REs in the TGRA have significant characteristics of a high frequency and low intensity (Gu et al., 2017; Deng et al., 2020a), which is obviously different from the macroseismic action. The rock joints only move in one direction under the cyclic shear load produced by frequent microseisms (Liu, 2017), i.e., the cyclic loading will not exceed the peak shear strength of the rock joints (only pre-peak cyclic loading and unloading occurs). Only a few studies have described the dynamic action of frequent microseisms on the rock joints as pre-peak cyclic shear loading (Li et al., 2003); based on this, Fathi et al. (2016), Kou et al. (2019) and Liu et al. (2020c) successively revealed the strength and deformation characteristics and the fatigue damage evolution process of the artificial rock joints using the pre-peak cyclic shear experiments. In addition, most of the existing literature only considers the influence of joint undulation on the fatigue damage mechanism of cyclic shear, and seldom considers the influence of joint roughness. Earlier, Patton (1966) used the first-order and second-order asperities to characterize the joint undulation and roughness, respectively, and Barton (1973), Hoek (1983) and Bandis (1983) showed that the cyclic shear fatigue damage characteristics of the rock joints are mainly controlled by the first-order asperities when the shear deformation or normal stress is large, otherwise, it is dominated by the second-order asperities. Based on this, Lee et al. (2001) and Liu et al. (2018b, 2020c) investigated the fatigue damage mechanical behaviours of the saw-tooth rock joints containing first-order and second-order asperities using cyclic shear experiments. In particular, the particle flow code 2D (PFC^{2D}) program has been adopted by most scholars because it can intuitively and accurately obtain the progressive fatigue damage evolution characteristics of crack initiation, propagation and penetration (asperity damage, local fracture and complete failure) of the rock joints during the entire shear process (Potyondy & Cundall, 2004; Asadi et al., 2012; Bahaaddini et al., 2013; Lambert & Coll, 2014; Liu et al., 2019). In summary, for complex loading conditions with the frequent occurrence of microseisms after the TGRA impoundment, a study on the fatigue damage mechanism of the rock joints under cyclic shear loading is more consistent with the real stress state of the rock joints on the reservoir bank slopes. Overall, most scholars use the laboratory cyclic shear experiments and the PFC^{2D} program as the general means and tools to investigate the fatigue damage mechanism of the rock joints. However, there are few reports on the fatigue damage mechanism of the rock joints under pre-peak cyclic shear loading, which take the physical model of natural rock processing as the research object and fully consider the influence of joint undulation and roughness.

In view of this, and based on previous studies (Liu, 2017; Liu et al., 2018b; Kou et al., 2019; Liu et al., 2019; Deng et al., 2020a; Liu et al., 2020c), the laboratory pre-peak cyclic shear experiments under a constant normal loading (CNL) were conducted, the shear deformation and strength characteristics of the rock joints with multiscale asperities were discussed, the influence rules of the normal stress, shear rate, shear amplitude and first-order asperity angle on the fatigue damage mechanism of the rock joints were analysed, and the macro fatigue damage evolution process and typical failure modes of the rock joints were revealed. Moreover, fully considering the number of cycles (increasing cycles) and joint morphology (including the wavy rock joints), the PFC^{2D} program was used to simulate the entire process of the pre-peak cyclic shear experiment, the variation rules of the meso fatigue damage crack number and energy of the rock joints were analysed, the meso fatigue damage evolution process and particle distribution characteristics of the rock joints were revealed, and the fatigue damage mechanism of the rock joints was generalized and discussed after comparing the macro-meso fatigue damage evolution process. The research results enrich the basic theory of the fatigue damage of the rock joints and provide reference value for dynamic stability analysis and evaluation of engineering rock masses.

2.2 Pre-Peak Cyclic Shear Experiment

2.2.1 Property Parameters of Limestone

The frequent occurrence of the REs in the TGRA since the impoundment in 2003 (Liu, 2018) has resulted in geological disasters of various types, with large-scale and severe hazards often occurring on the reservoir bank slopes (most of which are the BRSs) (Deng et al., 2020a). In particular, such slopes often contain the through-type saw-tooth rock joints (see Fig. 2.1a, the geographical base map was made using ArcGIS software), and the rock joints usually have large spatial extensibility and opening degree, causing most unstable sliding bodies to be densely developed in such slopes (Liu et al., 2018c).

The raw materials, i.e., the limestone derived from one landslide scene in the Wushan section of the TGRA (see Fig. 2.1b), were used in the experiments. The limestone was processed to a standard specimen to meet the experimental requirements (see Fig. 2.1c, its height $\times$ diameter $=$ 100 $\times$ 50 mm) using a rock cutting machine, a coring machine and a grinding machine. The mechanical parameters (see Table 2.1) of limestone were measured (see Fig. 2.1d) to provide the theoretical basis for the subsequent pre-peak cyclic shear experiments and the PFC^{2D} meso numerical simulations.

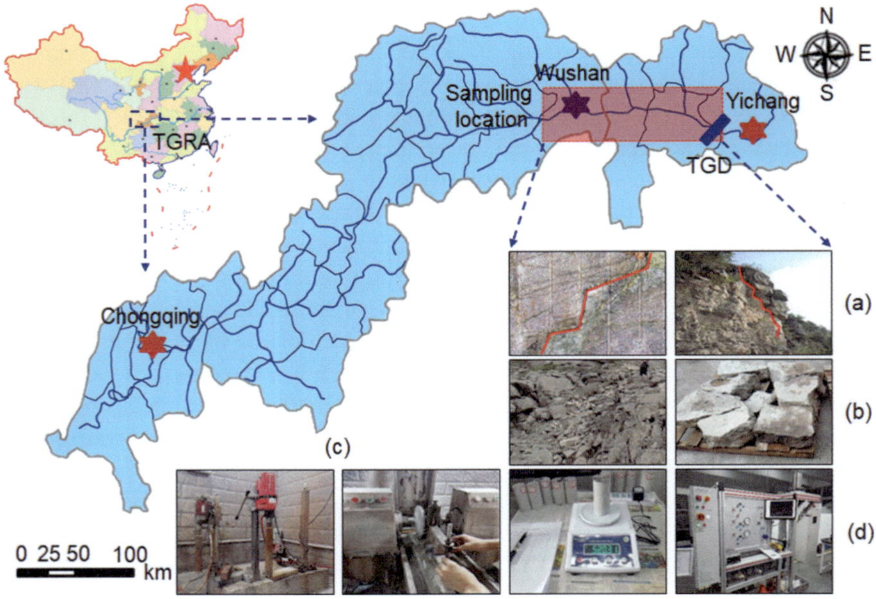

Fig. 2.1 **a** Slopes with the saw-tooth rock joints in the TGRA, **b** field sampling, **c** specimen preparation, and **d** determination of the mechanical parameters of limestone

Table 2.1 Physical–mechanical parameters of limestone

Density (g/cm^3)	Compressive strength (MPa)	Modulus of elasticity (MPa)	Poisson's ratio	Cohension (kPa)	Internal friction angle (°)
2.65	57.26	0.66×10^4	0.24	5260	44.53

2.2.2 Design of the Artificial Rock Joints

Existing studies (Liu, 2017; Gu et al., 2017; Liu et al., 2018c; Liu et al., 2019; Deng et al., 2020a) have noted that the earthquake force is equivalent to the cyclic load for the rock joints, and that the rock masses will have bidirectional dislocation along the joint surface when the earthquake force is large, which causes the joint surface to suffer periodic fatigue wear, and the shear strength parameters to significantly degraded. In contrast, the cyclic load on most rock joints will not exceed the peak shear strength of the joint surface, and the rock masses only move unidirectionally along the joint surface; in this case, the stress mode of the joint surface can be regarded as pre-peak cyclic loading and unloading. Taking the stress mode of a typical BRS with the through-type saw-tooth rock joints (red dotted line as the assumed main sliding surface) under a horizontal earthquake force as an example (see Fig. 2.2a). If the sliding mass moves upward along the main sliding surface, then the upward component force of the horizontal earthquake force along the main sliding surface

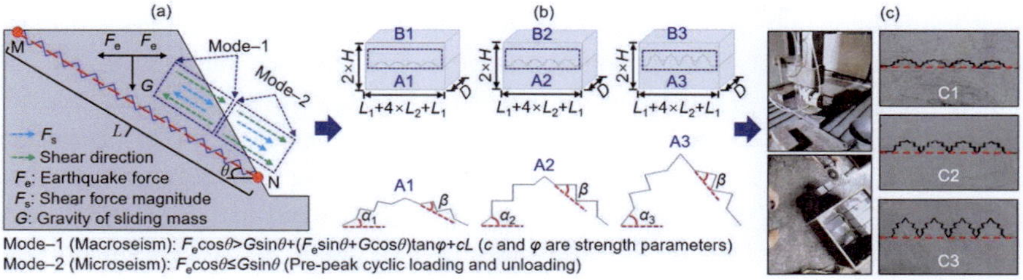

Fig. 2.2 **a** Simplified mechanical model of the saw-tooth rock joints under an earthquake force, **b** simplified configuration of the rock joints with multiscale asperities, and **c** processing of the artificial rock joints with multiscale asperities

needs to simultaneously overcome the downward component force of gravity of the sliding mass along the main sliding surface and the anti-sliding force on the main sliding surface, i.e., the horizontal earthquake force needs to be sufficiently large (macroseism) to make the sliding mass move upward along the main sliding surface, and then cause the sliding mass to move along the main sliding surface in a two-way cyclic shear dislocation (see Mode-1 in Fig. 2.2a). However, the horizontal earthquake force on the sliding mass under the RIS (microseism) is limited, and its upward component force along the main sliding surface usually does not exceed the component force of its self-weight downward along the main sliding surface, which makes it difficult for the sliding mass to move upward along the main sliding surface. At this time, the shear direction of the sliding mass is always downward along the main sliding surface during the earthquake action, and only the shear force magnitude changes periodically, i.e., the sliding mass only undergoes unidirectional cyclic loading and unloading along the main sliding surface (see Mode-2 in Fig. 2.2a).

Combined with Figs. 2.1a and 2.2a, in order to describe the geometric features of the rock joints more realistically, the joint geometry was generalized and designed, as given in models A1 ~ A3 (see Fig. 2.2b), after fully considering the rock strata undulation (represented by the first-order asperities) and roughness (represented by the second-order asperities). At the same time, according to the literature (Muralha et al., 2014) and fully considering the conditions of experimental equipment and specimen processing, the final geometric dimensions of models B1 ~ B3 were determined to be $L_1 = 30$ mm, $L_2 = 15$ mm, $H = D = 50$ mm, $\alpha_1 = 30°$, $\alpha_2 = 45°$, $\alpha_3 = 60°$ and $\beta = 45°$ (see Fig. 2.2b). In addition, a high-pressure water jet cutting machine and a rock grinding machine were used to process limestone into standard specimens C1 ~ C3 (see Fig. 2.2c) to meet the experimental requirements.

2.2.3 Working Condition Design

In the experiments, the effects of the normal stress, shear rate, shear amplitude, first-order asperity angle and number of cycles on the fatigue damage mechanism of the rock joints under pre-peak cyclic shear loading were fully considered, and then a total of nine groups of working conditions were designed (see Table 2.2).

Among them, the pre-peak cyclic shear experiments with 50, 100, 200, 500, 800 and 1000 cycles were successively conducted under each group of working condition. In particular, the average value of the peak shear strength (τ_p) of the rock joints under each group of working condition was obtained using the static direct shear experiments before performing the pre-peak cyclic shear experiments (see Table 2.3), which provides the upper and lower limit values of the shear load (each of them accounts for a certain proportion of the peak shear strength) for the cyclic shear stage, and the shear amplitude is the difference between the upper and lower limit value of the shear load (Liu, 2017; Liu et al., 2018b; Kou et al., 2019; Liu et al., 2020c).

Table 2.2 Working conditions of the pre-peak cyclic shear experiment

Working conditions	Surface shape of the rock joints	Normal stress (MPa)	Shear rate (kN/s)	Shear amplitude (%)	First-order (Second-order) asperity angle (°)	Number of cycles
1#	Saw-tooth	3.0	1.5	$50\tau_p$	45 (45)	50 100 200 500 800 1000
2#		1.5	1.5	$50\tau_p$	45 (45)	
3#		4.5	1.5	$50\tau_p$	45 (45)	
4#		3.0	0.5	$50\tau_p$	45 (45)	
5#		3.0	1.0	$50\tau_p$	45 (45)	
6#		3.0	1.5	$30\tau_p$	45 (45)	
7#		3.0	1.5	$40\tau_p$	45 (45)	
8#		3.0	1.5	$50\tau_p$	30 (45)	
9#		3.0	1.5	$50\tau_p$	60 (45)	

Table 2.3 Peak shear strength of the rock joints in the static direct shear experiment

Working conditions	Peak shear strength (MPa)				Average value (MPa)
1#	3.681	3.702	3.592	3.677	3.663
2#	2.504	2.633	2.522	2.565	2.556
3#	4.541	4.689	4.553	4.669	4.613
4#	4.167	4.112	3.985	4.144	4.102
5#	3.783	3.822	3.966	3.845	3.854
6#	3.977	4.225	4.003	3.943	4.037
7#	3.793	3.886	3.945	3.620	3.811
8#	3.297	3.502	3.436	3.305	3.385
9#	4.259	4.361	4.377	4.211	4.302

2.2.4 Equipment and Loading Scheme

The equipment used in the experiments mainly includes a loading system, a monitoring system, a data acquisition system and a shearing device (see Fig. 2.3a). The loading system is a WDAJ-600 type microcomputer controlled electro-hydraulic servo rock shear rheological testing machine, with a maximum axial (tangential) displacement stroke of 100 mm (the loading rate is 0.001 ~ 10 mm/min), and a maximum test force of 600 kN (the loading rate is 0.00167 ~ 1.67 kN/s). The monitoring system includes a high-speed camera and a computer, and the data acquisition system consists of a multi-channel intelligent data acquisition instrument and a computer. The shearing device is a visual, detachable and portable shear box that was independently developed and designed and can observe and photograph the macroscopic shear fatigue damage evolution process of the rock joints in real-time through a toughened glass window.

Combined with Fig. 2.2a, the specific loading scheme is illustrated by taking the shear process of working condition 1# under a CNL as an example. Before loading, a proper amount of lubricant was evenly applied on the contact surface of the upper and lower boxes to reduce its friction resistance, and the axial (tangential) rigid

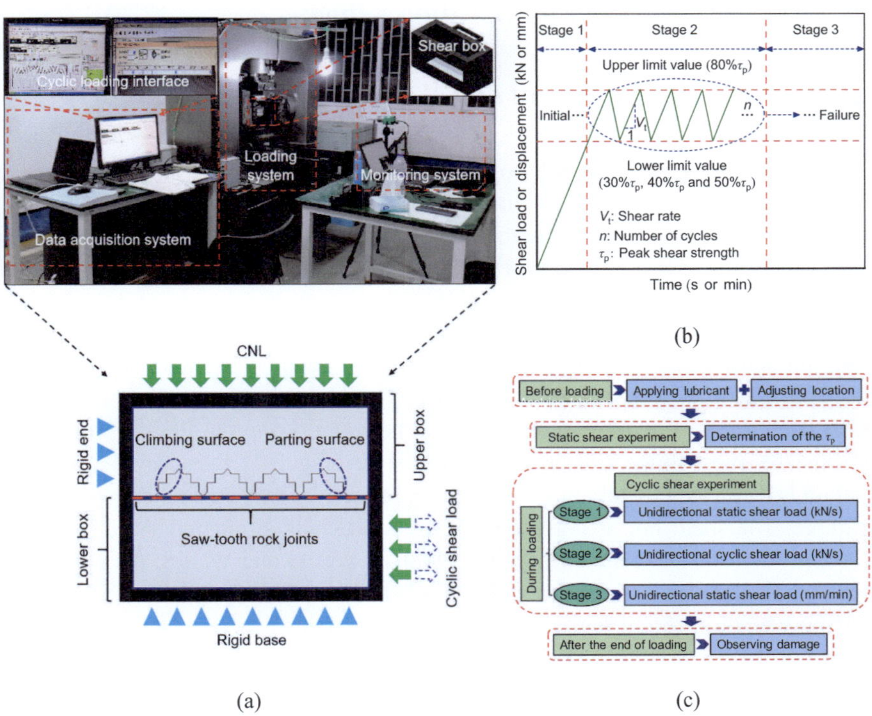

Fig. 2.3 **a** Experimental equipment and loading principle, **b** loading procedures, and **c** flowchart of the entire experimental process

loading head, rigid end and rigid base of the testing machine were aligned with the corresponding positions of the shear box (see Fig. 2.3a). During loading, first, a unidirectional static shear load was applied using the test force loading mode (the loading rate is 0.5, 1.0 or 1.5 kN/s) until it reached the lower limit value of the shear load in the cyclic shear stage (see Stage 1 in Fig. 2.3b, c). Second, the unidirectional cyclic shear load (loading and unloading) was applied using the test force loading mode (the loading rate is 0.5, 1.0 or 1.5 kN/s) (Liu, 2017; Liu et al., 2018b; Kou et al., 2019; Liu et al., 2020c) until the set number of cycles was completed (see Stage 2 in Fig. 2.3b, c). Finally, the unidirectional static shear load was applied using the displacement loading mode (the loading rate is 0.5 mm/min) until the rock joints were completely destroyed (see Stage 3 in Fig. 2.3b, c). After the end of loading, the upper and lower parts of the specimen were separated along the shear surface, and the phenomena of wear, passivation and fracture of the joint surface were observed and photographed (see Fig. 2.3a). In addition, the static direct shear experiments, which are used to determine the peak shear strength of the rock joints under each group of working condition, used the displacement loading mode (the loading rate is 0.5 mm/min) to apply a unidirectional static shear load.

2.3 PFC^{2D} Numerical Exploration Scheme

2.3.1 Model Establishment and Working Condition Design

To further study the variation rules of the fatigue damage crack number (and energy) and the meso fatigue damage evolution process (and particle distribution characteristics) of the rock joints from a meso perspective, four groups of working conditions of the meso numerical simulations (see Table 2.4), which fully consider the number of cycles (increasing cycles) and joint morphology (including the wavy rock joints), were designed based on the laboratory pre-peak cyclic shear experiments. Among them, the meso numerical simulations of pre-peak cyclic shear with 50, 100, 150, 200, 300, 400, 500, 600, 700, 800, 900 and 1000 cycles were successively conducted under each group of working condition. Furthermore, using the PFC^{2D} program to establish the meso numerical simulation model (see Fig. 2.4), the specific modelling method is illustrated by taking the modelling process of working condition #2 as an example. The shear box is composed of the rigid walls numbered Wall-1 ~ Wall-8, where the lower box includes Wall-1, Wall-2, Wall-3 and Wall-8, the upper box includes Wall-4, Wall-5, Wall-6 and Wall-7, and the Wall-3 and Wall-7 are used as wing walls to prevent particles from escaping when shearing (Liu et al., 2019). A total of 20546 circular particles with a porosity of 0.15 are randomly generated within the shear box, and the particle radius is uniformly distributed between 0.38 ~ 0.54 mm.

Additionally, the parallel bond model is used for the contact calculation of granular materials (Potyondy & Cundall, 2004; Lambert & Coll, 2014), the rock joints are

Table 2.4 Working conditions of the meso numerical simulation

Working conditions	Surface shape of the rock joints	Normal stress (MPa)	Shear rate (kN/s)	Shear amplitude (%)	First-order (Second-order) asperity angle (°)	Number of cycles
#1	Saw-tooth	3.0	1.5	$50\tau_p$	30 (45)	50 100 150 200 300 400 500 600 700 800 900 1000
#2					45 (45)	
#3					60 (45)	
#4	Wavy				45 (0)	

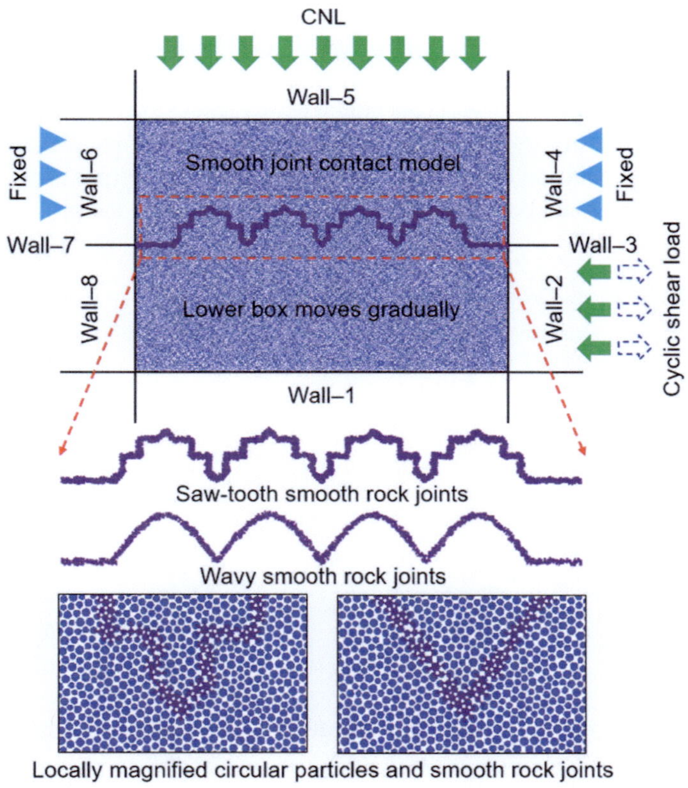

Fig. 2.4 Meso numerical simulation model established by the PFC^{2D} program

created by the discrete fracture net (Bahaaddini et al., 2013; Liu et al., 2020c), and the smooth joint contact model is given to the granular materials within the range of 0.6 mm on both sides of the rock joints (Asadi et al., 2012; Liu, 2017).

2.3.2 Parameter Calibration and Verification

The macro mechanical parameters obtained after calibrating (repeated trial calculations) the meso mechanical parameters of the meso numerical simulation via the trial and error method (Yoon, 2007) are shown in Table 2.5. The mechanical parameters in Tables 2.1 and 2.5 are in good agreement (the errors are less than $\pm$ 5%); thus, the meso mechanical parameters in Table 2.6 can be used for the meso numerical simulations of the particle flow. Figure 2.5 shows that the comparison of the macro-meso results of the shear stress—shear displacement curve and the failure morphology of the rock joints (including the wavy rock joints) in static direct shear under a CNL.

It can be seen from Fig. 2.5 that for the saw-tooth and wavy rock joints, although the shear stress—shear displacement curve obtained by the meso simulation has a significant fluctuation trend (especially in the post-peak stage) compared with that obtained from the macro experiment, the curves of the meso simulation and the macro experiment still show very consistent characteristics of phased development as a whole, and both can be divided into obvious elastic zone (pre-peak stress rise zone) and plastic zone (post-peak stress drop zone). Although the peak shear stress

Table 2.5 Macro parameters of the meso numerical simulation

Density (g/cm^3)	Compressive strength (MPa)	Modulus of elasticity (MPa)	Poisson's ratio	Cohension (kPa)	Internal friction angle (°)
2.65	56.49	0.67×10^4	0.24	5380	44.31

Table 2.6 Meso parameters of the meso numerical simulation

Circular particles		Smooth rock joints		Parallel bonds	
Frictional coefficient	0.6	Bonding mode	1	Radius multiplier	1.0
Mass density (kg/m^3)	2750	Dilation angle (°)	0	Shear strength (MPa)	62.5
Maximum radius (mm)	0.54	Frictional coefficient	0.5	Normal strength (MPa)	48.5
Modulus of elasticity (GPa)	12	Tensile strength (MPa)	0	Modulus of elasticity (GPa)	25
Ratio of normal to shear stiffness	1.5	Shear stiffness (GPa/m)	80	Ratio of normal to shear stiffness	2.5
Ratio of maximum to minimum radius	1.42	Normal stiffness (GPa/m)	80	SD of shear and normal strength (MPa)	$\pm$ 5

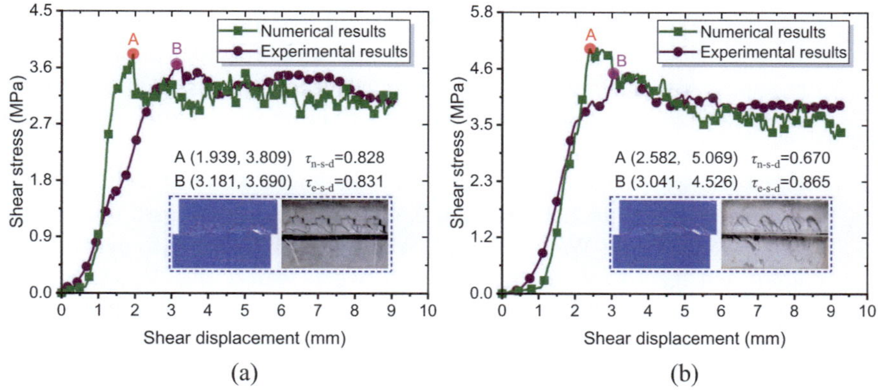

Fig. 2.5 Comparison of the meso numerical results and the macro experimental results: **a** saw-tooth rock joints, and **b** wavy rock joints; $\tau_{n\text{-}s\text{-}d}$ and $\tau_{e\text{-}s\text{-}d}$ are the stress drops of the meso numerical simulation and the macro laboratory experiment, respectively

of the rock joints obtained by the meso simulation is larger than that of the macro experiment, while the peak shear displacement and stress drop ($\tau_{n\text{-}s\text{-}d}$ and $\tau_{e\text{-}s\text{-}d}$) are smaller than those of the macro experiment, the peak shear stress, peak shear displacement and stress drop obtained from the meso simulation and the macro experiment are still generally consistent. Overall, the failure patterns obtained by the meso simulation are in good agreement with those obtained from the macro experiment, with both revealing the phenomena of crack initiation, propagation and penetration, as well as the extrusion, wear, void, fracture and migration of the first-order and second-order asperities. Therefore, based on the above calibrated meso mechanical parameters (see Table 2.6), it is reasonable and feasible to study the fatigue damage mechanical behaviours of the saw-tooth and wavy rock joints under pre-peak cyclic shear loading by the PFC^{2D} program.

2.3.3 Realization Method of the Pre-Peak Cyclic Shear Process

Based on the servo loading principle of the PFC^{2D} program (Liu et al., 2018b), the servo function program is compiled using the FISH language to dynamically control the real-time rate of the rigid walls of shear box, and finally realize the application process of constant normal load and pre-peak cyclic shear load.

(1) Realization method of constant normal loading: Combined with Fig. 2.5, the servo function program is called to dynamically update the real-time vertical rate of Wall-5, and to monitor and record its normal stress (the Wall-4, Wall-6 and Wall-7 are set as fixed boundaries). When the ratio of the difference between the normal stress and the target stress to the target stress is less than

0.3%, the servo loading process of the normal load on Wall-5 is considered to be completed (Liu et al., 2019). Then, the servo function program is run continuously (gradually eliminating the unbalanced force inside the specimen) until the statically determinate specimen with a uniform bearing constant normal load is obtained.

(2) Realization method of pre-peak cyclic shear loading: Combined with Figs. 2.3b and 2.5, the servo function program is called to set the forward quasi-static horizontal motions of Wall-1, Wall-2, Wall-3 and Wall-8, and to dynamically monitor and record the real-time force between Wall-2 and Wall-8 in order to calculate the internal shear stress of the specimen according to the force equilibrium equation. When the shear stress reaches the upper limit value of the pre-peak cyclic shear stress, the backward quasi-static horizontal motions of Wall-1, Wall-2, Wall-3 and Wall-8 are set (Liu, 2017). Then, the internal shear stress of the specimen is continuously monitored and recorded, and when the shear stress reaches the lower limit value of the pre-peak cyclic shear stress, the forward quasi-static horizontal motions of Wall-1, Wall-2, Wall-3 and Wall-8 are set. In this way, those move forward and backward repeatedly until the target number of cycles is completed. Finally, the Wall-1, Wall-2, Wall-3 and Wall-8 are set to move forward quasi-static and horizontally until the end of the entire loading process.

2.4 Experimental Results and Analysis

2.4.1 Shear Deformation and Strength Characteristics

Figure 2.6 show the relationship curves between the shear stress and the shear displacement of the rock joints under different numbers of cycles (taking working condition 1# as an example). Figure 2.7 show the relationship curves between the shear stress and the shear displacement of the rock joints under the same number of cycles (taking 1000 cycles as an example). It can be seen from Figs. 2.6 and 2.7 that the relationship curve between the shear stress and the shear displacement of the rock joints under pre-peak cyclic shear loading can be roughly divided into six stages.

(1) Stage 1: The rock joints are in a state of initial nonlinear shear contraction deformation, where the shear stress—shear displacement curve shows an approximate variation trend of the upward concave shape (the growth rate is very small). The effective contact area between the asperities on both sides of the joint surface increases gradually, and is accompanied by a slight normal moving component (shear contraction). At this stage, the rock joints, which are in the full elastic deformation zone, present the effect of compacting damage.

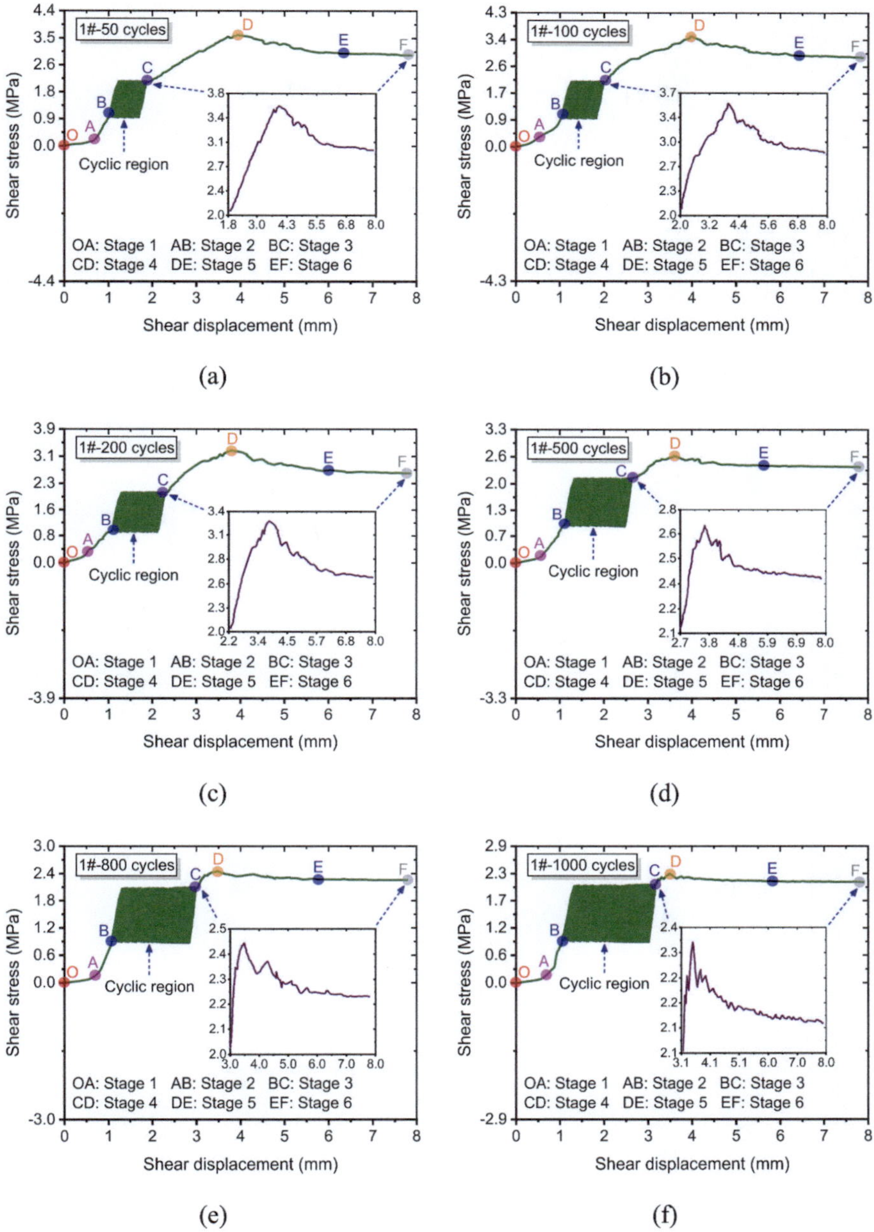

Fig. 2.6 Shear stress versus shear displacement of the rock joints: **a**–**f** represent the test results of 50, 100, 200, 500, 800 and 1000 cycles under working condition 1#

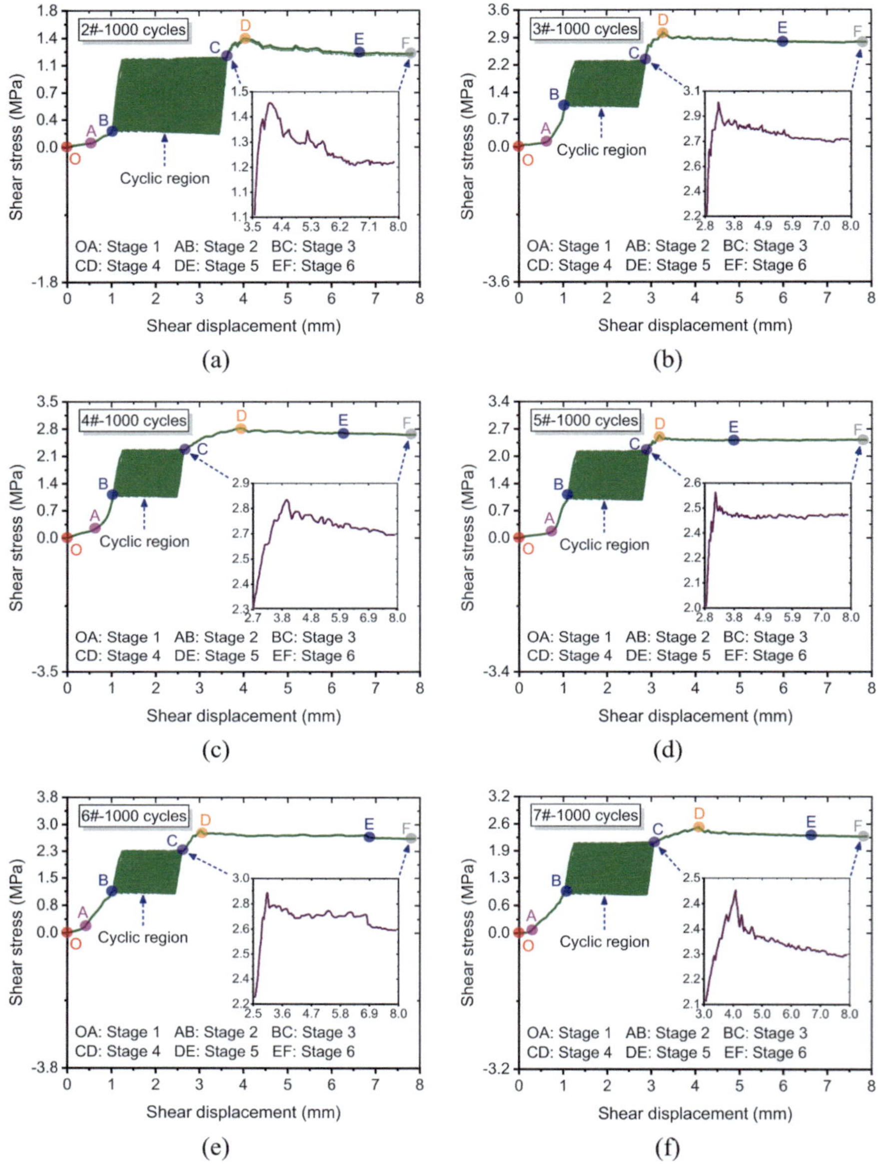

Fig. 2.7 Shear stress versus shear displacement of the rock joints: **a**–**h** represent the test results of 1000 cycles under working conditions 2# ~ 9#

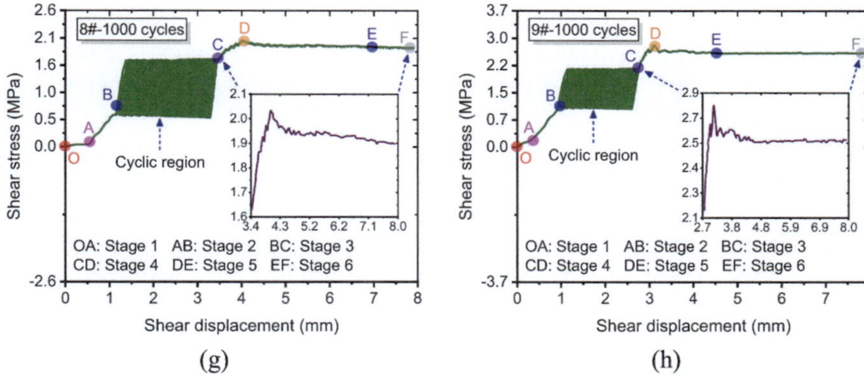

Fig. 2.7 (continued)

(2) Stage 2: The rock joints are in a state of approximate linear elastic shear dilation deformation, where the shear stress—shear displacement curve shows an approximate linear variation trend (the curve slope essentially remains unchanged). The asperities on the upper side of the joint surface move obliquely upward along the climbing surfaces (see Fig. 2.3a, similarly hereinafter) of the first-order asperities, and are accompanied by a slight normal moving component (shear dilation). At this stage, the rock joints, which are in the approximate linear elastic deformation zone, present the effect of climbing damage.

(3) Stage 3: The rock joints are in a state of cyclic fatigue damage deformation, where the shear stress—shear displacement curve shows an approximate variation trend of the hysteresis loop shape (the shear stress increases and decreases periodically). The wear scars between the asperities on both sides of the joint surface are gradually obvious, and the cumulative normal shear dilation deformation gradually increases. At this stage, the rock joints, which are in the elastic–plastic deformation zone, present the effect of climbing—cyclic abrading damage.

(4) Stage 4: The rock joints are in a state of plastic deformation of the local compression-shear fracture, where the shear stress—shear displacement curve can be divided into approximately linear and nonlinear sections. The asperities on the upper side of the joint surface tend to climb over the tip of the first-order asperities (the normal dilation deformation continuously increases with an approximately linear feature), and the second-order asperities of the rock joints experience an instantaneous local fracture (the curve exhibits a mutation point). At this stage, the rock joints, which are in the plastic deformation zone, present the effect of extruding—gnawing damage.

(5) Stage 5: The rock joints are in a state of full plastic deformation of the stress brittle drop, where the shear stress—shear displacement curve has the characteristics of multi-peaks (post-peak stress and deformation are extremely irregular, and the peaks and valleys obviously alternate fluctuating). Oblique fracture cracks occur in the middle and upper parts of the first-order asperities of the

rock joints, which cause an instantaneous brittle drop in the shear stress and slow down the growth rate of normal shear dilation deformation. At this stage, the rock joints, which are in the full plastic deformation zone, present the effect of gnawing damage.

(6) Stage 6: The rock joints are in a state of ideal plastic flow deformation, where the shear stress—shear displacement curve shows an approximate linear fluctuation variation trend. The first-order asperities of the rock joints gradually slide along the fracture surface after they are completely fractured along their root parts (the stick–slip friction is significant and the residual deformation process is characterized by an ideal plastic friction sliding), and the leading factors of the shear mechanical behaviours of the joint surface have changed. At this stage, the rock joints, which are in the ideal plastic deformation zone, present the effect of sliding damage. In particular, with an increase in the number of cycles, the peak (and residual) shear strengths of the rock joints decrease gradually, while the cumulative shear (and normal) deformations increase gradually, and the rock joint strength reaches its peak value faster (the slope of the shear stress—shear displacement curve is relatively large).

2.4.2 *Factors Influencing the Fatigue Damage Mechanism*

2.4.2.1 Normal Stress

Figure 2.8a, b show the curves of the peak (and residual) shear strengths and the cumulative shear (and normal) displacements of the rock joints after undergoing pre-peak cyclic shear loading versus the number of cycles under different normal stresses, respectively. It can be seen from Fig. 2.8 that the peak (and residual) shear strength and the cumulative shear (and normal) displacement of the rock joints under the same normal stress gradually decrease and increase, respectively, with an increase in the number of cycles.

The curve of the peak (and residual) shear strengths versus the number of cycles can be roughly divided into three stages (see Fig. 2.8a), i.e., the stress approximately shows a change trend (downward concave shape) of a slight decrease—sharp decrease (Stage 1), the stress approximately shows a linear change trend of a sharp decrease (Stage 2), and the stress approximately shows a linear change trend of a slow decrease (Stage 3).The first two stages take roughly 200 cycles as the boundary value, while the boundary value of the last two stages is approximately 500 cycles.

The curve of the cumulative shear (and normal) displacements versus the number of cycles can be roughly divided into two stages (see Fig. 2.8b), i.e., the displacement approximately shows a change trend (downward concave shape) of a sharp increase—slow increase (Stage 1) and the displacement approximately shows a linear change trend of a slow increase (Stage 2), where the boundary value of these two stages is approximately 200 cycles. Moreover, the peak (and residual) shear strengths and the cumulative shear (and normal) displacements of the rock joints under the same

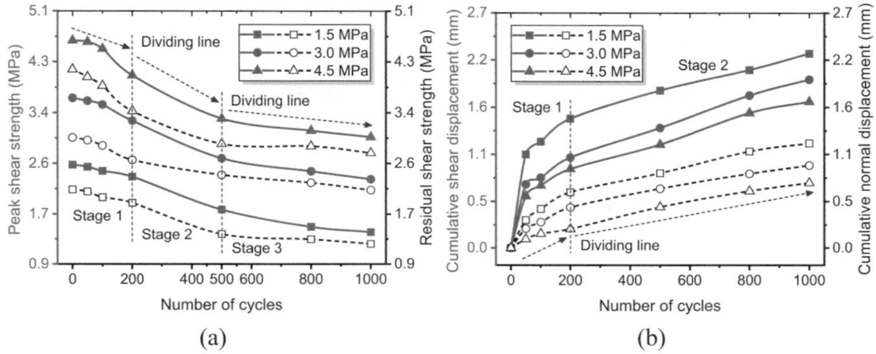

Fig. 2.8 Influence of the normal stress on the fatigue damage mechanism of the rock joints: **a** peak (and residual) shear strengths versus the number of cycles, and **b** cumulative shear (and normal) displacements versus the number of cycles

number of cycles gradually increase and decrease, respectively, with an increasing normal stress. As the normal stress increases (1.5, 3.0 and 4.5 MPa), the peak (and residual) shear strengths of the rock joints decreased by 43.82% (42.36%), 36.75% (28.97%) and 34.60% (33.40%), respectively, from 0 to 1000 cycles; while the cumulative shear (and normal) displacements increased by 51.77% (73.16%), 62.04% (76.39%) and 64.31% (86.30%), respectively, from 50 to 1000 cycles.

It can be seen from this that when the normal stress is small, the rock joints under cyclic shear loading are prone to developing a large shear and normal deformation (the periodic dislocation and expansion are obvious), and the deformation degree becomes increasingly significant with an increase in the number of cycles. When the normal stress is large, the rock joints under cyclic shear loading are prone to developing abrading, extruding and gnawing damage effects, and the damage effects (progressive fatigue damage process) gradually enhance with an increase in the number of cycles.

2.4.2.2 Shear Rate

Figure 2.9a, b show the curves of the peak (and residual) shear strengths and the cumulative shear (and normal) displacements of the rock joints after undergoing pre-peak cyclic shear loading versus the number of cycles under different shear rates, respectively. It can be seen from Fig. 2.9 that the variation rules of the peak (and residual) shear strengths and the cumulative shear (and normal) displacements with the number of cycles (development trend, stage division and partition boundary value of the curve) are essentially consistent with the above under the same shear rate (see Sect. 2.4.2.1).

Moreover, the peak (and residual) shear strengths and the cumulative shear (and normal) displacements of the rock joints under the same number of cycles gradually decrease and increase, respectively, with an increasing shear rate. As the shear rate increases (0.5, 1.0 and 1.5 kN/s), the peak (and residual) shear strengths of the

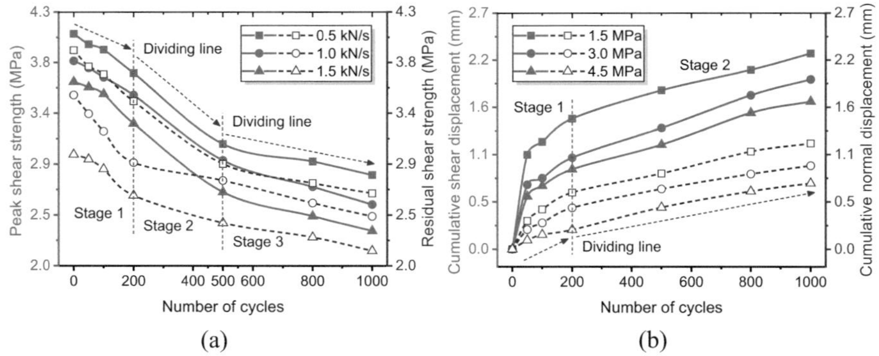

Fig. 2.9 Influence of the shear rate on the fatigue damage mechanism of the rock joints: **a** peak (and residual) shear strengths versus the number of cycles, and **b** cumulative shear (and normal) displacements versus the number of cycles

rock joints decreased by 31.23% (32.82%), 33.76% (30.95%) and 36.75% (28.97%), respectively, from 0 to 1000 cycles; while the cumulative shear (and normal) displacements increased by 70.87% (89.23%), 68.17% (78.17%) and 62.04% (76.39%), respectively, from 50 to 1000 cycles. It can be seen from this that when the shear rate is small, the horizontal compression between the asperities on both sides of the joint surface is not obvious, which results in the development rate of the rock joints' fatigue damage slowing down and the strength degradation degree weakening. When the shear rate is large, the climbing surfaces of the first-order asperities rapidly develop abrading and extruding damage effects (deformation develops rapidly), which leads to the critical number of cycles of the rock joints' fatigue damage being reduced and the strength degradation degree being intensified.

2.4.2.3 Shear Amplitude

Figure 2.10a, b show the curves of the peak (and residual) shear strengths and the cumulative shear (and normal) displacements of the rock joints after undergoing pre-peak cyclic shear loading versus the number of cycles under different shear amplitudes, respectively. It can be seen from Fig. 2.10 that the variation rules of the peak (and residual) shear strengths and the cumulative shear (and normal) displacements with the number of cycles (development trend, stage division and partition boundary value of the curve) are essentially consistent with the above under the same shear amplitude (see Sect. 2.4.2.1).

Moreover, the peak (and residual) shear strengths and the cumulative shear (and normal) displacements of the rock joints under the same number of cycles gradually decrease and increase, respectively, with an increasing shear amplitude. As the shear amplitude increases ($30\tau_p\%$, $40\tau_p\%$ and $50\tau_p\%$), the peak (and residual) shear strengths of the rock joints decreased by 29.08% (28.31%), 34.72% (31.84%) and 36.75% (28.97%), respectively, from 0 to 1000 cycles; while the cumulative shear

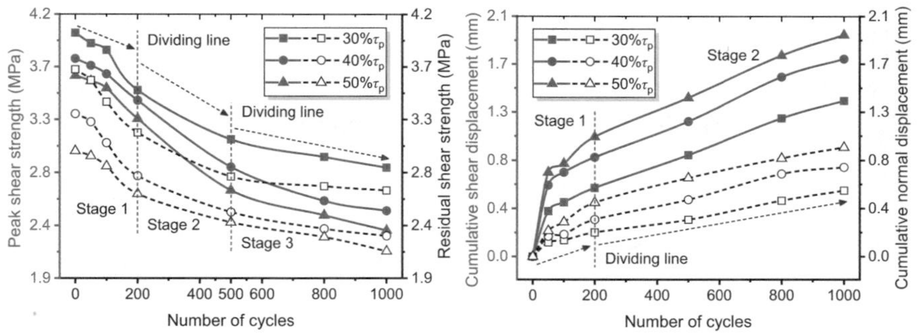

(a) (b)

Fig. 2.10 Influence of the shear amplitude on the fatigue damage mechanism of the rock joints: **a** peak (and residual) shear strengths versus the number of cycles, and **b** cumulative shear (and normal) displacements versus the number of cycles

(and normal) displacements increased by 70.77% (78.47%), 63.94% (77.28%) and 62.04% (76.39%), respectively, from 50 to 1000 cycles. It can be seen from this that when the shear amplitude is small, the cyclic shear process of the rock joints is relatively gentle, and the abrading damage effect caused by the local mutual extrusion and the shear between asperities on both sides of the joint surface is weak, which slows down the fatigue damage rate and weakens the strength degradation degree of the rock joints. Additionally, when the shear amplitude is large, the significant horizontal compression damage occurs rapidly between the asperities on both sides of the joint surface, which leads to an increase in the strength deterioration degree and the deformation development speed of the rock joints.

2.4.2.4 First-Order Asperity Angle

Figure 2.11a, b show the curves of the peak (and residual) shear strengths and the cumulative shear (and normal) displacements of the rock joints after undergoing pre-peak cyclic shear loading versus the number of cycles under different first-order asperity angles, respectively. It can be seen from Fig. 2.11 that the variation rules of the peak (and residual) shear strengths and the cumulative shear (and normal) displacements with the number of cycles (development trend, stage division and partition boundary value of the curve) are essentially consistent with the above under the same first-order asperity angle (see Sect. 2.4.2.1). Moreover, the peak (and residual) shear strengths and the cumulative shear (and normal) displacements of the rock joints under the same number of cycles gradually increase and decrease, respectively, with an increasing first-order asperity angle. As the first-order asperity angle increases (30°, 40°and 50°), the peak (and residual) shear strengths of the rock joints decreased by 39.94% (34.15%), 36.75% (28.97%) and 35.52% (25.64%), respectively, from 0 to 1000 cycles; while the cumulative shear (and normal) displacements increased by 59.79% (62.82%), 62.04% (76.39%) and 66.96% (80.53%), respectively, from

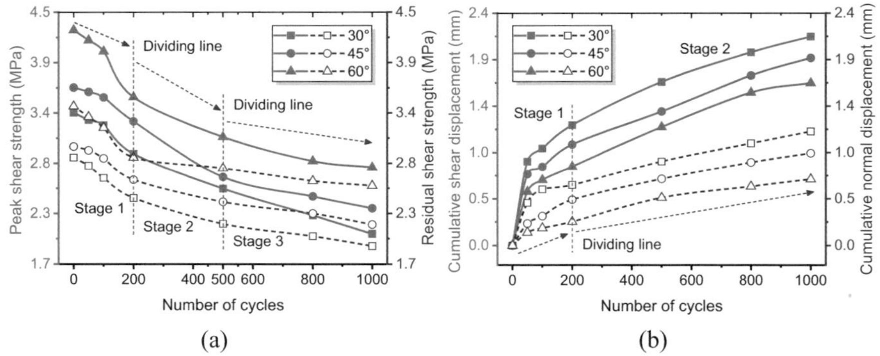

Fig. 2.11 Influence of the first-order asperity angle on the fatigue damage mechanism of the rock joints: **a** peak (and residual) shear strengths versus the number of cycles, and **b** cumulative shear (and normal) displacements versus the number of cycles

50 to 1000 cycles. It can be seen from this that when the first-order asperity angle is small, the rock joints easily experience climbing and climbing—gnawing damage (the shear and normal deformations are usually large), and the strength degradation degree is relatively large. When the first-order asperity angle is large, the shear strength performance of the rock joints is significantly affected by the asperities, and the gnawing damage effect occurs rapidly between the asperities on both sides of the joint surface (the strength degradation generally has the characteristics of an instantaneous drop).

2.4.3 Macro Fatigue Damage Evolution Process and Typical Failure Modes

Figure 2.12 shows the macro fatigue damage evolution process of the rock joints under different numbers of cycles (taking working condition 1# as an example). Figure 2.13 shows the macro fatigue damage evolution process of the rock joints under the same number of cycles (taking 1000 cycles as an example). It can be seen from Fig. 2.12 that the macro fatigue damage phenomena (crack initiation, propagation and penetration, as well as the extrusion, wear, void, fracture and migration of the first-order and second-order asperities) of the rock joints under pre-peak cyclic shear loading are gradually significant with an increase in the number of cycles. Combined with Figs. 2.6 and 2.7, It can be seen from Fig. 2.13 that the macro fatigue damage evolution process of the rock joints under pre-peak cyclic shear loading can be roughly divided into three stages.

(1) Stage P1: The joint surface is closed by the combined action of an initial compression-shear, resulting in slight shear contraction deformation. Subsequently, the asperities on the upper side of the joint surface move obliquely

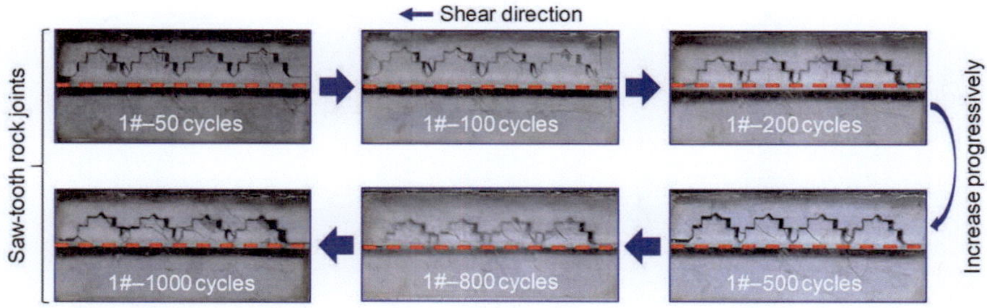

Fig. 2.12 Macro fatigue damage evolution process and failure morphology of the rock joints under pre-peak cyclic shear loading: the experimental results of 50, 100, 200, 500, 800 and 1000 cycles under working condition 1#

upward along the climbing surfaces of the first-order asperities and produce slight shear dilation deformation. Moreover, the meshing degree between the asperities on both sides of the joint surface increases, resulting in the initiation of cracks on the climbing surfaces of the first-order asperities, and a slight gap appears between the parting surfaces (see Fig. 2.3a, similarly hereinafter). At this stage, the rock joints show a compacting—climbing failure mode (corresponding to the OB section in Figs. 2.6 and 2.7).

(2) Stage P2: The asperities on the upper side of the joint surface have the deformation trend of climbing over the tip of the first-order asperities, and the shear dilation deformation increases significantly. Under cyclic shear loading, the wear scars between the asperities on both sides of the joint surface are gradually obvious, and the cumulative shear dilation deformation steadily increases. Subsequently, the local fracture of the second-order asperities occurs under a concentrated compressive stress, and the gap between the parting surfaces increases significantly. Moreover, there is a gradual increase in the cracks on the climbing surfaces of the first-order asperities, and a few of them expand and extend obliquely downward from the middle and upper parts of the climbing surfaces. At this stage, the rock joints show a climbing—cyclic abrading—compacting—gnawing failure mode (corresponding to the BD section in Figs. 2.6 and 2.7).

(3) Stage P3: Under the combined action of a continuous compression-shear, the number of oblique cracks on the climbing surfaces of the first-order asperities increases sharply, and all the first-order asperities rapidly and completely fracture brittlely and obliquely from the middle and upper parts of the climbing surfaces, while the root parts of the first-order asperities approximately completely penetrate along the horizontal direction. The fracture sequence of the first-order asperities occurs one by one from the rigid end of the testing machine to the loading end of the shear load. Subsequently, the asperities on the upper side of the joint surface climb over the tip of the first-order asperities, and then exhibit significant stick–slip friction along the fracture surface, with the increasing rate of shear dilation deformation essentially remaining unchanged.

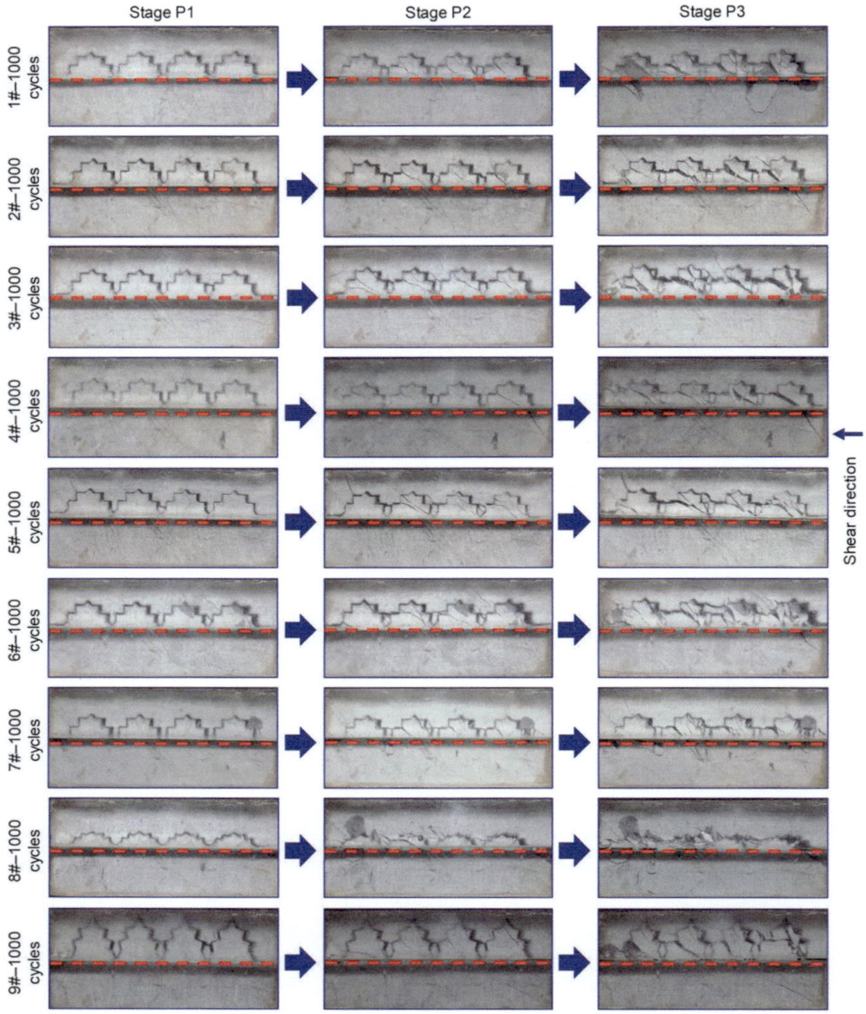

Fig. 2.13 Macro fatigue damage evolution process and failure morphology of the rock joints under pre-peak cyclic shear loading: the experimental results of 1000 cycles under working conditions 1# ~ 9#

Moreover, a large number of loose rock blocks, particles and other debris, which are produced by compaction, climbing, cyclic abrading, extrusion and gnawing of multiscale asperities, continuously dislocate, rub and migrate. Finally, the shear mechanical behaviours of the joint surface change gradually, from being controlled by the first-order and second-order asperities, to being dominated by the mechanical properties of loose rock blocks and particles. At this stage, the rock joints show a gnawing—sliding failure mode (corresponding to the DF section in Figs. 2.6 and 2.7). In particular, when the normal stress, shear rate, shear amplitude and first-order asperity angle are large, the compacting, contacting, abrading and gnawing damage effects of the rock joints are obviously enhanced with an increasing number of cycles, which makes the ultimate fatigue damage degree of the rock joints increases significantly.

2.5 Numerical Exploration Results and Analysis

2.5.1 *Variation Rules of the Meso Crack Number and Energy*

Figure 2.14 shows the relationship curves between the meso fatigue damage crack number (and energy) and the shear displacement of the rock joints (including the wavy rock joints) under pre-peak cyclic shear loading (taking 1000 cycles as an example), while Fig. 2.15 shows the relationship curves of that with the number of cycles. In particular, the meso fatigue damage crack includes tensile damage crack and shear damage crack. The meso fatigue damage energy is the external work of Wall-2 caused by shear loading in the meso particle flow simulation, which includes tensile shear dissipation energy (dissipation energy of the tensile shear damage on the surface and inside of the rock joints) and elastic strain energy (releasable elastic strain energy stored in the rock masses).

It can be seen from Fig. 2.14 that the meso fatigue damage crack number (and energy) of the rock joints show an obvious stage growth trend with an increase in the shear displacement, and the relationship curves between the meso fatigue damage crack number (and energy) and the shear displacement can be roughly divided into three development stages. (1) Stage 1: Slight increase—sharp increase (early stage), where the meso fatigue damage crack number (and energy) approximately show a

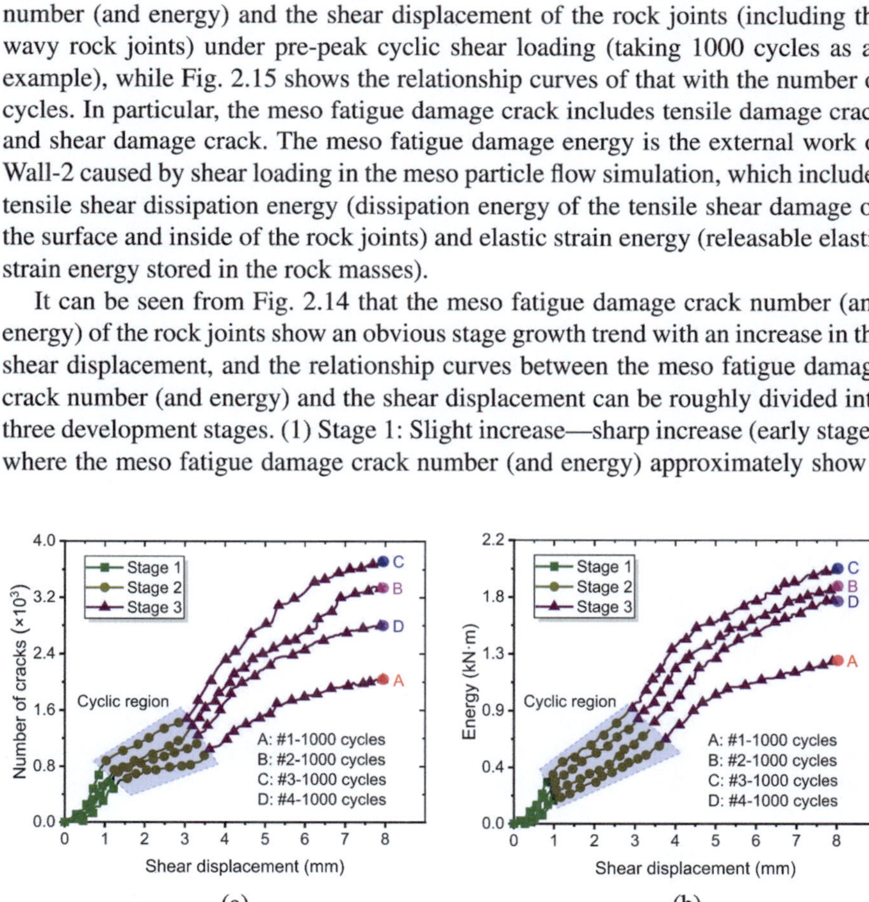

Fig. 2.14 **a** Number of cracks in the meso fatigue damage versus the shear displacement of the rock joints, and **b** energy in the meso fatigue damage versus the shear displacement of the rock joints

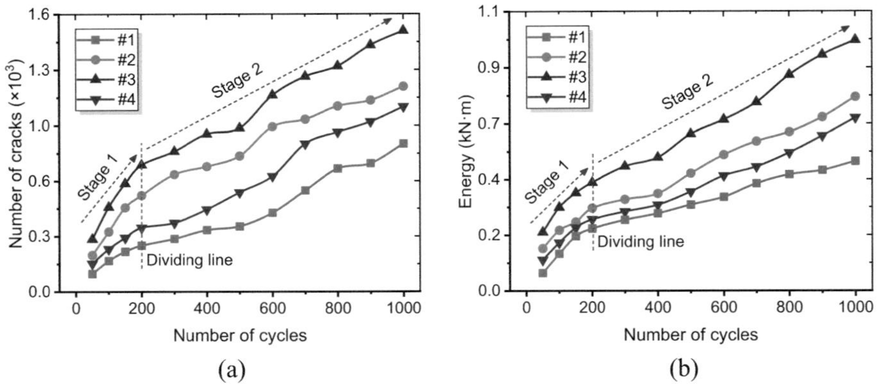

Fig. 2.15 **a** Number of cracks in the meso fatigue damage versus the number of cycles of the rock joints, and **b** energy in the meso fatigue damage versus the number of cycles of the rock joints

growth trend of the upward concave shape, namely, their increase rates are small at first and then large. At this stage, the rock joints have compacting and climbing damage effects. (2) Stage 2 (cyclic region): Slow increase (middle stage), where the meso fatigue damage crack number (and energy) approximately show a linear growth trend, namely, their increase rates are essentially unchanged. At this stage, the rock joints have cyclic abrading damage effect. (3) Stage 3: Sharp increase—slow increase—slight increase (later stage), where the meso fatigue damage crack number (and energy) first approximately show a growth trend of the downward concave shape, and finally approximate to a horizontal linear growth trend, namely, their increase rates are large at first and then small, until it essentially remains unchanged. At this time, the rock joints have extruding, gnawing and sliding damage effects. In addition, under the same shear displacement, the meso fatigue damage crack number (and energy) of the saw-tooth rock joints increase with an increase in the first-order asperity angle, and the meso fatigue damage crack number (and energy) of the wavy rock joints are less than those of the saw-tooth rock joints.

It can be seen from Fig. 2.15 that the meso fatigue damage crack number (and energy) of the rock joints show an obvious stage growth trend with an increase in the number of cycles, and the relationship curves between the meso fatigue damage crack number (and energy) and the number of cycles can be roughly divided into two development stages. (1) Stage 1: Sharp increase—slow increase (early stage), where the meso fatigue damage crack number (and energy) approximately show a growth trend with downward concave shape, namely, the increase rates of those are large at first and then small. (2) Stage 2: Slow increase (later stage), where the meso fatigue damage crack number (and energy) approximately grow in a linear trend, and the volatilities of those are obvious. In particular, these two stages take roughly 200 cycles as the boundary value. Additionally, under the same number of cycles, the meso fatigue damage crack number (and energy) of the saw-tooth rock joints increase with an increase in the first-order asperity angle, and the meso fatigue damage crack number (and energy) of the wavy rock joints are less than those of the saw-tooth rock joints.

2.5.2 Meso Fatigue Damage Evolution Process and Particle Distribution Features

Figure 2.16 shows the meso fatigue damage evolution process of the rock joints (including the wavy rock joints) under different numbers of cycles (taking working conditions #2 and #4 as an example). Figure 2.17 shows the meso fatigue damage evolution process and the distribution characteristics (rose diagram) of particle fatigue damage (intergranular bond fracture) of the rock joints under the same number of cycles (taking 1000 cycles as an example).

It can be seen from Fig. 2.16 that with an increase in the number of cycles, the number of meso fatigue damage particles of the rock joints under pre-peak cyclic shear loading increases gradually. Moreover, when the number of cycles is small, the meso fatigue damage particles mainly occurs on the climbing surfaces of the first-order asperities. In contrast, the meso fatigue damage particles continuously expand and extend in an oblique direction from the middle and upper parts of the climbing surfaces. Finally, the meso fatigue damage particles penetrate the root parts of the first-order asperities, and the meso cracks on the parting surfaces of the first-order asperities expand continuously.

Combined with Figs. 2.6, 2.7, 2.12 and 2.13, it can be seen from Fig. 2.17 that the meso fatigue damage evolution process of the rock joints under pre-peak cyclic shear loading can be roughly divided into three development stages.

(1) Stage 1: The contact effect between the meso particles on both sides of the joint surface is strengthened under the combined action of an initial compression-shear (compacting damage effect). Then, the meso particles on the climbing surfaces of the first-order asperities are gradually damaged under continuous loading, including pinnate meso cracks initiated by shear friction and meso

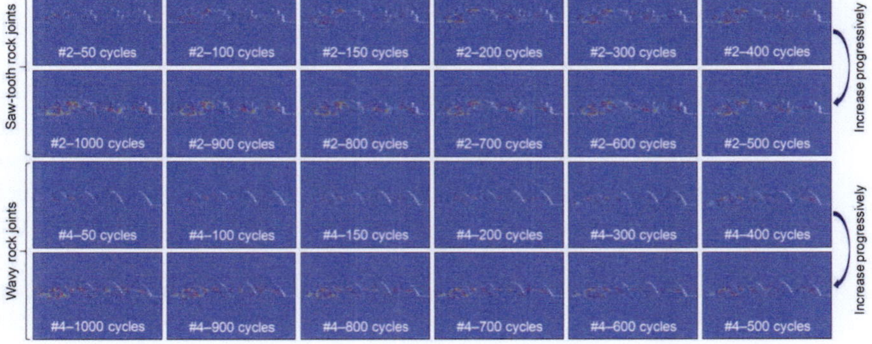

Fig. 2.16 Meso fatigue damage evolution process and failure morphology of the rock joints under pre-peak cyclic shear loading: the numerical results of 50, 100, 150, 200, 300, 400, 500, 600, 700, 800, 900 and 1000 cycles under working conditions #2 and #4

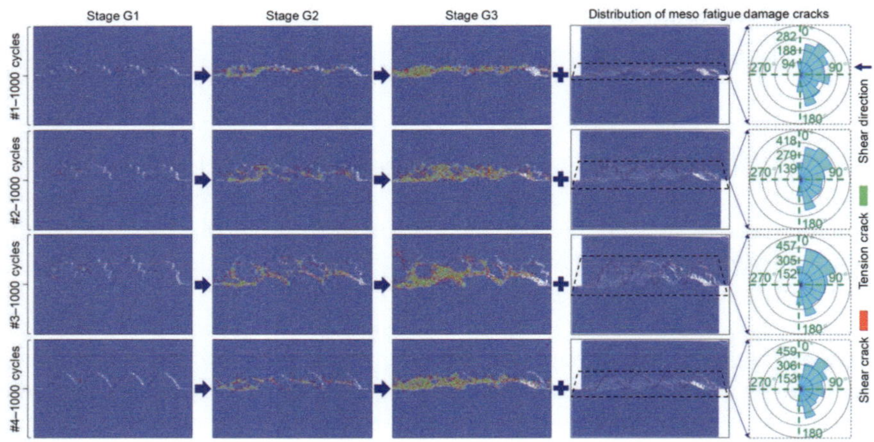

Fig. 2.17 Meso fatigue damage evolution process and failure morphology of the rock joints under pre-peak cyclic shear loading: the numerical results of 1000 cycles under working conditions #1 ~ #4

scratches caused by sliding friction. At this time, the meso fissures appear near the parting surfaces of the first-order asperities (climbing damage effect).

(2) Stage 2: The damage degree of the meso particles on the climbing surfaces of the first-order asperities is more serious and has an obvious vertical movement trend, and the friction rolling effect between the meso particles on both sides of the joint surface is enhanced under cyclic shear loading (climbing—cyclic abrading damage effect). Then, the meso fatigue damage particles are densely distributed near the second-order asperities, and there are several oblique meso tensile cracks that almost penetrate in the middle and upper parts of the climbing surfaces of the first-order asperities, while there are also meso shear cracks at its tip. At this time, the meso fissures near the parting surfaces of the first-order asperities are further expanded (extruding—gnawing damage effect).

(3) Stage 3: Under the combined action of a continuous compression-shear, the meshing degree between the meso particles on both sides of the joint surface is significantly enhanced, until several damage paths of the meso particles on the climbing surfaces of the first-order asperities are completely connected and their root parts are completely fractured. At this time, the meso fissures near the parting surfaces of the first-order asperities are filled with a large number of meso fatigue damage particles, and the boundary of the failure region of the rock joints is significantly expanded (gnawing damage effect). Then the meso fatigue damage particles gradually migrate along the through-type joint surface (mutual friction), and their distribution range continues to expand to both sides of the joint surface (sliding damage effect).

In general, the number of meso fatigue damage particles increases, and the failure region expands with an increasing first-order asperity. Influenced by the mechanical effect of the second-order asperities of the saw-tooth rock joints, the number of meso fatigue damage particles is greater than that of the wavy rock joints. However, the final fatigue damage distribution regions of the meso particles both of them are similar under the same first-order asperity angle. In addition, the meso fatigue damage particles are densely distributed near the joint surface in an approximately trapezoidal manner under pre-peak cyclic shear loading, and the inclination angle of the contact force between the particles is between 0° and 360° when the meso particles are damaged (see the rose diagram on the right side in Fig. 2.17). Among them, the maximum number of the fatigue damage particles appears in the range of 0° to 90°, followed by 90° to 180°, with very few appearing in the range of 180° to 270° and 270° to 0°. In particular, the comparative analysis of Figs. 2.6, 2.7, 2.12, 2.13, 2.16 and 2.17 shows that the macro-meso fatigue damage evolution process and failure modes (mechanical response) of the rock joints under pre-peak cyclic shear loading are in good agreement.

2.5.3 Generalized Description of the Macro-Meso Fatigue Damage Mechanism

Combined with Figs. 2.6, 2.7, 2.12, 2.13, 2.14, 2.15, 2.16 and 2.17, the macro-meso fatigue damage mechanism (evolution process and failure modes) of the rock joints (including the wavy rock joints) under pre-peak cyclic shear loading can be generally described using three typical development stages, as shown in Fig. 2.18, i.e., stages P1 (G1) ~ P3 (G3), which corresponds to stages P1 ~ P3 in Figs. 2.12 and 2.13, and stages G1 ~ G3 in Figs. 2.16 and 2.17, where the values of N_C in Fig. 2.18 are 50, 100, 150, 200, 300, 400, 500, 600, 700, 800, 900 and 1000.

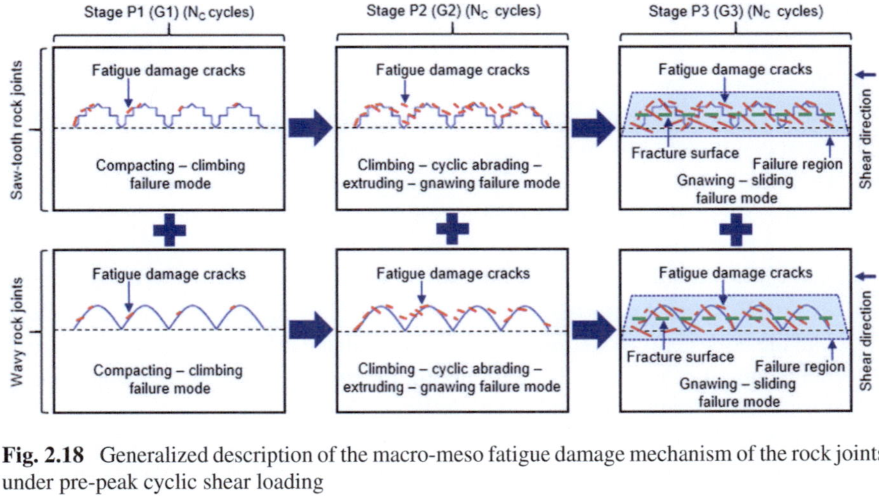

Fig. 2.18 Generalized description of the macro-meso fatigue damage mechanism of the rock joints under pre-peak cyclic shear loading

(1) Stage P1 (G1): The rock joints exhibit compacting damage effect (initial nonlinear shear contraction deformation) and climbing damage effect (approximate linear elastic dilatancy deformation). At this time, the number of the fatigue damage cracks is very small, and the cracks are mainly distributed on the climbing surfaces. Meanwhile, the fatigue damage energy is small. In this stage, the rock joints are generally in the elastic deformation zone and mainly show the compacting—climbing failure mode.
(2) Stage P2 (G2): The rock joints exhibit climbing—cyclic abrading damage effect (cyclic fatigue damage deformation) and extruding—gnawing damage effect (plastic deformation of the local compression-shear fracture). At this time, the number of the fatigue damage cracks increases gradually, and the length and spatial expansion of a single crack obviously increase (the connectivity of the fracture path tends to be complete). Meanwhile, the fatigue damage energy increases sharply. In this stage, the rock joints are generally in the elastic–plastic deformation zone and mainly show the climbing—cyclic abrading—extruding—gnawing failure mode.
(3) Stage P3 (G3): The rock joints exhibit gnawing damage effect (full plastic deformation of the stress brittle drop) and sliding damage effect (ideal plastic flow deformation). At this time, the number of the fatigue damage cracks increases significantly, with the cracks densely distributed in the vicinity of the joint surface in an approximately trapezoidal manner and form a regular failure region, and a fracture surface that approximately runs through the middle parts of the first-order asperities. Meanwhile, the increment of the fatigue damage energy tends to be stable. In this stage, the rock joints are generally in the plastic deformation zone and mainly show the gnawing—sliding failure mode.

2.6 Summary and Conclusions

(1) The pre-peak cyclic shear deformation and strength characteristics of the rock joints show that the change process of the shear stress with the shear displacement can be roughly divided into six development stages, i.e., initial nonlinear shear contraction deformation (compacting damage effect), approximate linear elastic shear dilation deformation (climbing damage effect), cyclic fatigue damage deformation (climbing—cyclic abrading damage effect), plastic deformation of the local compression-shear fracture (extruding—gnawing damage effect), full plastic deformation of the stress brittle drop (gnawing damage effect) and ideal plastic flow deformation (sliding damage effect), corresponding to the deformation zone of full elasticity, approximate linear elasticity, elastoplasticity, plasticity, full plasticity and ideal plasticity, respectively.
(2) Under the same normal stress, shear rate, shear amplitude or first-order asperity angle, the peak (and residual) shear strengths and cumulative shear (and normal) displacements of the rock joints decrease and increase, respectively, with an increase in the number of cycles. Under the same number of cycles, the peak (and

residual) shear strengths and the cumulative shear (and normal) displacements of the rock joints gradually increase and decrease, respectively, when the normal stress or first-order asperity angle increases; while they respectively decrease and increase when the shear rate or shear amplitude increases. The change process of the peak (and residual) shear strengths with the number of cycles can be roughly divided into three development stages, i.e., stress slight decline—sharp decline, stress approximate linear sharp decline and stress approximate linear slow decline, where the first two stages take roughly 200 cycles as the boundary value, while the boundary value of the last two stages is approximately 500 cycles. The change process of the cumulative shear (and normal) displacementa with the number of cycles can be roughly divided into two development stages, i.e., displacement sharp increase—slow increase and displacement approximate linear slow increase, where the boundary value of these two stages is approximately 200 cycles.

(3) The variation process of the meso fatigue damage crack number (and energy) with the shear displacement of the rock joints subjected to pre-peak cyclic shear loading can be approximately divided into three stages, i.e., slight increase—sharp increase (compacting and climbing damage effect), slow increase (cyclic abrading damage effect) and sharp increase—slow increase—slight increase (extruding, gnawing and sliding damage effect) in the early stage, middle stage and later stage, respectively. While the change process of that with the number of cycles includes two stages, i.e., sharp increase—slow increase and slow increase in the early stage and later stage, respectively, with these two stages take roughly 200 cycles as the boundary value.

(4) The PFC^{2D} meso numerical results are in good agreement with the macro experimental results by comparing and analysing the macro-meso fatigue damage evolution process (the failure morphologies of multiscale asperities and the distribution characteristics of damage particles) of the rock joints. Furthermore, the macro-meso fatigue damage mechanism of the rock joints with multiscale asperities is generallly described, and the macro-meso fatigue damage failure modes of the rock joints under pre-peak cyclic shear loading can be summarized as three basic types, i.e., compacting—climbing failure mode, climbing—cyclic abrading—extruding—gnawing failure mode and gnawing—sliding failure mode. The meso fatigue damage cracks of the rock joints are densely distributed in the vicinity of the joint surface in an approximately trapezoidal manner, and the maximum number of the meso fatigue damage cracks appears when the inclination angle of the contact force between the particles is between 0° and 90°.

Chapter 3
New Algorithm for the Dynamic Stability of the BRS Considering Strength Degradation of the Rock Joints

3.1 Introduction

Generally, when analyzing and evaluating the stability of slope engineering, it is necessary to consider the impact of seismic load on it, and exploring the dynamic stability of slope has always been a very important hot issue in the field of geotechnical and seismic engineering (Liu et al., 2018c; Kasama et al., 2021). In particular, on the one hand, because the seismic load is a dynamic load, the slope will be subjected to dynamic seismic inertial force (SIF) during the seismic duration (Liu, 2017; Pang et al., 2021); on the other hand, because the seismic load is a cyclic shear load, the shear strength of the rock joints will continue to vibrate and degrade during the seismic duration (Liu et al., 2014; Liu et al., 2018d); that is, the dynamic stability coefficient (D_{SC}) of slope during the seismic duration is a real-time dynamic variable under the joint action of the above two factors. It can be seen that how to embed the strength vibration degradation principle of the rock joints into the dynamic stability analysis of slope, then, to quickly and accurately evaluate the stability of slope under dynamic load has important theoretical and practical significance.

At present, many achievements have been made in the research of the dynamic stability analysis methods of slope, and various analysis methods have been continuously developed, innovated and applied in engineering practice. The existing typical dynamic stability analysis methods of the rock slope mainly include the pseudo-static method (Siad, 2003; Nagpal & Basha, 2012), numerical analysis method (based on the limit equilibrium theory (LET)) (Zhang & Chen, 2001; Cai et al., 2002; Cheng et al., 2007; Stianson et al., 2015; Liu et al., 2015; Li et al., 2021), strength reduction theory (Zheng et al., 2010; Tu et al., 2016; Sun et al., 2020) or numerical stress field theory (Liu, 2010) and pseudo-dynamic method (Steedman & Zeng, 1990; Choudhury & Nimbalkar, 2006; Ghosh, 2007). However, the above analysis methods have not taken into account the irreversible degradation effect of cyclic dynamic load due to seismic action on the shear strength properties of the rock joints. In fact, a large number of dynamic tests have shown that the shear mechanical behaviors of the

X. Liu and B. Xu, *Dynamic Stability of the Bedding Rock Slope with Deteriorated Hydro-fluctuation Belt*, https://doi.org/10.1007/978-981-95-6311-1_3

rock joints subjected to cyclic dynamic load will have dynamic damage degradation effect. Especially, Liu et al. (2014), Liu et al. (2018b), Kou et al. (2019) and Xu et al. (2021) systematically analyzed the degradation rules of the shear strength of the rock joints with the number of cycles, cyclic shear amplitude and cyclic shear rate via the cyclic shear test of the artificial rock joints; Lee et al. (2001), Homand et al. (2001), Jafari et al. (2003) and Liu et al. (2011) emphatically investigated the degradation rules of the shear strength of the rock joints with the continuous action of cyclic shear load, and its damage evolution characteristics using the dynamic cyclic shear test of the natural or artificial rock joints. Although the above studies have obtained several preliminary conclusions, most of them only stay in the exploratory stage of the dynamic test of indoor rock samples, and have not been applied to the dynamic analysis of actual rock mass. Earlier, Crawford and Curran (1981), and Wang and Zhang (1982) found the friction factor of the rock joints under the moving state is a dynamic variable that changes with the accumulated displacement (vibration wear is gradually obvious) and relative motion velocity via the dynamic test; thus, the algorithm of the seismic permanent displacement of slope is established, and the evaluation principle of the dynamic stability of slope is preliminarily discussed. On this basis, Qi (2007), Ni et al. (2013) and Xia et al. (2021) successively revised and innovated the above research results, and proposed a algorithm of the seismic permanent displacement of slope considering the strength vibration degradation effect of the rock joints; then, they initially tried to apply it to the dynamic analysis of actual rock mass, and thus laying a theoretical and practical foundation for the analysis and evaluation the dynamic analysis of slope.

In conclusion, it can be seen that the vibration degradation of the rock joints during the seismic duration will lead to the dynamic attenuation evolution of its shear strength characteristic parameters; and the dynamic cumulative damage (DCD) of the strength of the rock joints after a certain period of dynamic time will certainly have an irreversible adverse impact on the final stability of slope. However, there are still few reports on the research on the dynamic stability of slope considering the strength vibration degradation effect of the rock joints at home and abroad. Therefore, how to quantitatively describe the strength vibration degradation effect of the rock joints, and apply it to the dynamic stability analysis of the rock slope is a scientific problem that needs to be further explored.

In view of this, the analytical solution of the strength vibration degradation coefficient (SVDC) of the rock joints affected by multiple factors (vibration wear and relative motion velocity between rock blocks) was firstly obtained based on the dynamic test. Then, the analytical solution was embedded into the equation for solving the D_{SC} of the rock slope. Subsequently, considering the strength vibration degradation effect of the rock joints, the dynamic algorithm of the D_{SC} of slope was realized by using the built-in FISH language programming of the universal distinct element code (UDEC) software that as the calculation engine. That is, the strength characteristic parameters of the rock joints at any moment of seismic duration were refreshed in real-time in each calculation time-step, and the SIF of slope at that moment is captured by the dynamic vector calculation of the net nodal force of grid node (Ni et al., 2013); furthermore, the time history curve of the D_{SC} of slope was obtained by dynamic calculation in combination with the LET, and thus the minimum strength vibration degradation coefficient ($SVDC_{min}$) (Liu et al., 2003) was used to solve the value of the final dynamic stability evaluation index of slope. In particular, the simple harmonic wave was used to simulate the seismic load, and the response characteristics of the dynamic stability of the rock slope with the penetrating straight rock joints were analyzed by using the algorithm proposed, i.e., the necessity of considering the strength vibration degradation effect of the rock joints when solving the D_{SC} of slope was clarified. Meanwhile, the algorithm was applied to the case study of the dynamic stability of slope under the background of Wenchuan seismic wave, the analysis results are consistent with the shaking table model test results, which further verifies the feasibility of the algorithm. Based on this, the response rules of the dynamic stability of slope under the influence of different strength characteristic parameters the rock joints (initial cohesion and internal friction angle), geometric parameters of slope (slope angle and dip angle of plane) and dynamic characteristic parameters (seismic load amplitude and seismic load frequency) were explored. Finally, the method of solving the instantaneous D_{SC} of the rock slope with the penetrating saw-tooth rock joints was discussed, and the current research achievements and future prospects were expounded, which is expected to provide a reliable theoretical basis for the long-term dynamic stability analysis and evaluation of the rock slope.

3.2 Analytical Solution of the SVDC

3.2.1 Strength Degradation Equation and Its Influencing Factors

Generally, under the action of seismic load (especially cyclic shear action), the shear strength of the rock joints (reflected by cohesion and internal friction angle) will produce degradation effect, which is accompanied by the entire process of complex dynamic load action. To quantitatively describe the dynamic degradation effect of the shear strength of the rock joints subjected to cyclic shear action of seismic load, a SVDC that changes dynamically with seismic time was introduced to describe this cumulative damage effect. Furthermore, combined with the classical Mohr–Coulomb strength criterion (Barton, 1973; Hoek and Bray, 1981), the shear strength vibration degradation equation of the rock joints at any moment of seismic duration can be obtained, as shown in Eqs. (3.1) and (3.2). In Eqs. (3.1) and (3.2): at any moment t of seismic duration, $\tau(t)$ is the shear strength of the rock joints, $\sigma(t)$ is the normal stress acting on the rock joints, $c(t)$ is the cohesion of the rock joints, $\varphi(t)$ is the internal friction angle of the rock joints, and $D_{sc}(t)$ is the SVDC of the rock joints; in addition, τ_0 is the initial shear strength of the rock joints, c_0 is the initial cohesion of the rock joints, φ_0 is the initial internal friction angle of the rock joints, and α is the asperity angle of the rock joints.

$$\tau(t) = \sigma(t)\tan\varphi(t) + c(t) = \tau_0 D_{sc}(t) = \sigma(t)\tan[\varphi_0 D_{sc}(t)] + c_0 D_{sc}(t), \alpha = 0 \tag{3.1}$$

$$\begin{aligned}\tau(t) &= \sigma(t)\tan\varphi(t) + c(t) = \tau_0 D_{sc}(t) \\ &= \sigma(t)\tan[(\varphi_0 + \alpha)D_{sc}(t)] + c_0 D_{sc}(t), \alpha \neq 0\end{aligned} \tag{3.2}$$

Based on this, the equation of the shear strength characteristic parameters of the rock joints at any moment of seismic duration can be obtained, as shown in Eqs. (3.3)–(3.5). In Eqs. (3.3)–(3.5): the physical meanings of all symbols are the same as those above-mentioned. Therefore, it can be seen from Eqs. (3.1)–(3.5) that after calculating the SVDC of the rock joints, the shear strength characteristic parameters of the rock joints at any moment of seismic duration can be obtained. Furthermore, according to the LET, dynamic vector method (DVM) and force polygon method (FPM) (Goodman, 1984; Zhang & Chen, 2001; Cai et al., 2002; Cheng et al., 2007; Ni et al., 2013; Stianson et al., 2015; Liu et al., 2015; Li et al., 2021), the dynamic change relationship of the D_{SC} of slope with the seismic time can be obtained.

$$\varphi(t) = \arctan\{\tan[\varphi_0 D_{sc}(t)]\}, \alpha = 0 \tag{3.3}$$

$$\varphi(t) = \arctan\{\tan[(\varphi_0 + \alpha)D_{sc}(t)]\}, \alpha \neq 0 \tag{3.4}$$

$$c(t) = c_0 D_{sc}(t) \tag{3.5}$$

At present, in the existing studies, the influencing factors of the cyclic shear action of seismic load on the degradation of the shear strength properties of the rock joints are generally summarized as vibration wear and relative motion velocity. On the one hand, the cumulative damage phenomena such as wear and passivation will occur gradually under the cyclic shear action of seismic load, resulting in irreversible attenuation of the shear strength of the rock joints. Earlier, several scholars used the cumulative displacement (C_D) of rock blocks to evaluate the influencing factors of vibration wear (Crawford & Curran, 1981; Wang & Zhang, 1982); and later, several scholars successively used the number of cycles and cyclic shear amplitude to evaluate the influencing factors of vibration wear (Lee et al., 2001; Ni et al., 2013; Xia et al., 2021). On the other hand, the relative motion velocity between rock blocks of slope under the cyclic shear action of seismic load will reduce the friction coefficient of the rock joints, i.e., there is a certain functional relationship between the dynamic friction coefficient of rock block and relative motion velocity. Earlier scholars revealed the dynamic change rules of the dynamic friction coefficient of rock block (under the movement state) with the relative motion velocity via the dynamic test, and established the dynamic differential equation of rock block of slope; that is, there is an even function relationship between the dynamic friction coefficient of rock block and absolute value of the relative motion velocity, and the former increases when the later decreases (Wang & Zhang, 1982). Subsequently, several scholars found that the dynamic friction coefficient of rock block under sliding state is a fluctuating variable, and its value is smaller than the static friction coefficient via the dynamic test of rigid block (Wang et al., 1994). It is found that the wear degradation of the rock joints is not only related to the number of cycles, but also closely related to the cyclic shear amplitude via the cyclic shear test of the rock joints. In addition, this study will use the above-mentioned functional relationship between the dynamic friction coefficient and relative motion velocity of rock blocks for reference, to describe the influence of the relative motion velocity on the degradation of the shear strength characteristic parameters of the rock joints.

3.2.2 Equation of the SVDC Affected by Multiple Factors

In view of the above analysis, this study describes the impact of the vibration wear and relative motion velocity on the shear strength characteristic parameters of the rock joints by using the influence factor of vibration wear and the influence factor of relative motion velocity, respectively. When calculating the SVDC of the rock joints, it is assumed that the influence factors of vibration wear and relative motion velocity independently during the entire seismic duration. Based on this, the equation of the SVDC (affected by multiple factors) of the rock joints at any moment of seismic duration can be obtained, as shown in Eq. (3.6). In Eq. (3.6): at any moment t of seismic duration, $\xi(t)$ is the influence factor of vibration wear of the rock joints, and $\lambda(t)$ is the influence factor of relative motion velocity of the rock joints; in addition, the physical meanings of other symbols are the same as those above-mentioned.

$$D_{sc}(t) = \xi(t)\lambda(t) \tag{3.6}$$

3.2.2.1 Equation for Solving the Influence Factor of Vibration Wear

The original limestone samples collected from a landslide site in the TGRA area were processed into standard specimens meeting the test requirements by cutting and grinding (as shown in Fig. 2.1), and further the cyclic shear tests were carried out on those specimens. Among them, the number of cycles includes 50, 100, 200, 500, 800 and 1000, the cyclic shear amplitude includes $30\%\tau_p$, $40\%\tau_p$ and $50\%\tau_p$; additionally, the normal stress is 3.0 MPa, and the cyclic shear rate is 1.0 kN/s. After each working condition is completed, the ratio of the peak shear strength of the rock joints after undergoing the set number of cycles to the peak shear strength of the rock joints without cyclic shear is calculated, and this ratio is defined as the degradation ratio of shear strength (DR_S) of the rock joints. Thus, the relationship curves between the DR_S of the rock joints and the number of cycles under different cyclic shear amplitudes are drawn, as shown in Fig. 3.1.

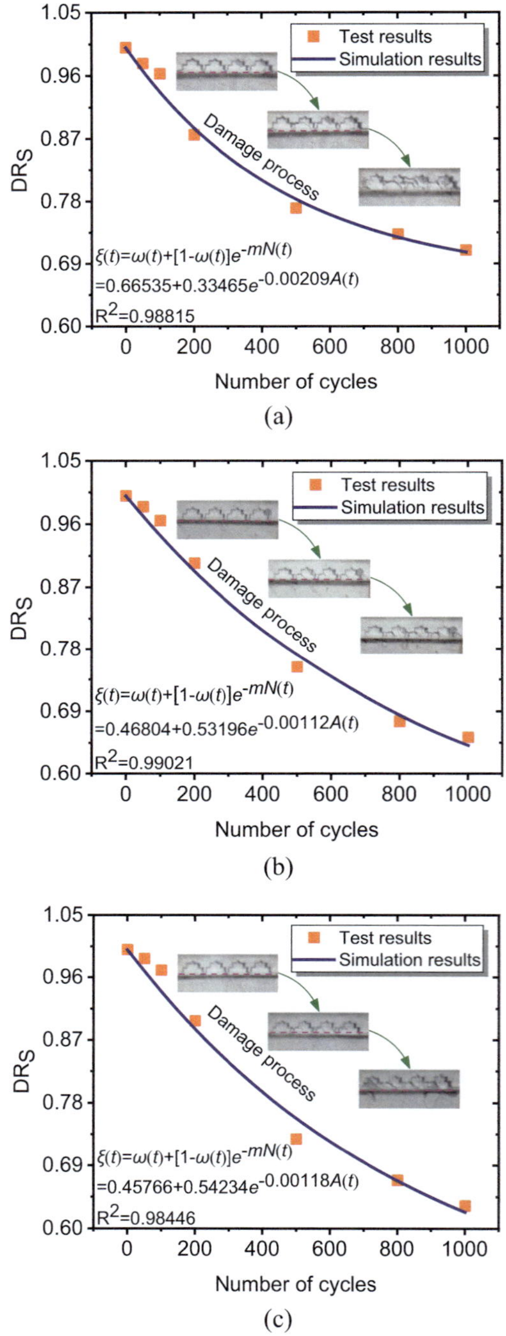

Fig. 3.1 Number of cycles versus DR_S of the rock joints subjected to cyclic shear loading: the shear amplitudes of (**a**), (**b**) and (**c**) are 30%, 40% and 50%, respectively, where τ_p is the average value of static peak shear strength of the rock joints (see Sect. 3.2.2.1 for the specific meanings of the symbols in this figure)

It can be seen from Fig. 3.1 that the test results of each group reflect the degradation rules of the shear strength of the rock joints well, i.e., the DR_S of the rock joints decreases with the increase of the number of cycles. The shear strength decays rapidly in a few shear cycles in the early stage, and after a certain number of cycles, the shear strength basically remains stable (the decay trend is slow). In general, there is a negative exponential decay relationship between the number of cycles and the DR_S of the rock joints under different cyclic shear amplitudes. Additionally, the attenuation degree of the DR_S of the rock joints is related to the cyclic shear amplitude, i.e., the greater the cyclic shear amplitude, the stronger the attenuation degree, and the smaller the final convergence value. It can be seen that the wear degradation of the rock joints is closely related to the number of cycles and cyclic shear amplitude, and a negative exponential function converging to $\omega(t)$ can be used to describe the influence factor of vibration wear of the rock joints, and its equation is shown in Eq. (3.7). In Eq. (3.7): $\omega(t)$ is the convergence value of the influence factor of vibration wear of the rock joints, m is the undetermined coefficient, and $N(t)$ is the number of cycles of the rock joints at any moment t of seismic duration; in addition, the physical meanings of other symbols are the same as those above-mentioned.

$$\xi(t) = \omega(t) + [1 - \omega(t)]e^{-mN(t)} \tag{3.7}$$

It can also be seen from Fig. 3.1 that when the number of cycles reach around 500, the DR_S of the rock joints under each cyclic shear amplitude is no longer obvious, and it can be considered that the shear strength of the rock joints has converged after 1000 cycles. The relationship curve between the DR_S of the rock joints and the cyclic shear amplitude after 1000 cycles is shown in Fig. 3.2 (reflecting the convergence value of the influence factor of vibration wear). It can be seen from Fig. 3.2 that the DR_S of the rock joints decreases with the increase of cyclic shear amplitude, and the two basically follow the negative exponential decay relationship; therefore, it can be assumed that the equation of the convergence value of the influence factor of vibration wear of the rock joints is shown in Eq. (3.8). In Eq. (3.8): P_0 is the convergence value of the influence factor of vibration wear of the rock joints, n is the undetermined coefficient, and $A(t)$ is the cyclic shear amplitude of the rock joints at any moment t of seismic duration; in addition, the physical meanings of other symbols are the same as those above-mentioned. Therefore, combined with Eqs. (3.7) and (3.8), the equation for solving the influence factor of vibration wear can be obtained, as shown in Eq. (3.9), and the physical meanings of all symbols in Eq. (3.9) are the same as those above-mentioned. In particular, combined with Figs. 3.1 and 3.2, $P_0 = 0.49386$, $m = 0.00146$ and $n = 0.719$ in Eq. (3.9) can be determined by regression analysis of least square method.

$$\omega(t) = P_0 + (1 - P_0)e^{-nA(t)} \tag{3.8}$$

$$\xi(t) = P_0 + (1 - P_0)e^{-nA(t)} + (1 - P_0)\left[1 - e^{-nA(t)}\right]e^{-mN(t)} \tag{3.9}$$

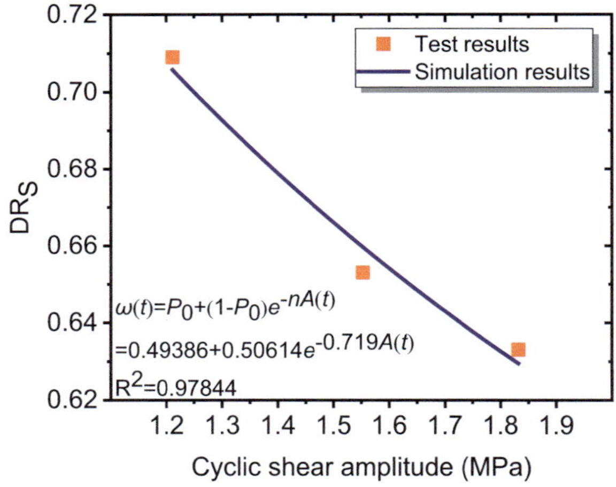

Fig. 3.2 Cyclic shear amplitude versus DR_S of the rock joints subjected to cyclic shear loading (see Sect. 3.2.2.1 for the specific meanings of the symbols in this figure)

3.2.2.2 Equation for Solving the Influence Factor of Relative Motion Velocity

The research results of the dynamic test obtained by Wang and Zhang (1982) were cited when solving the equation of the influence factor of relative motion velocity of the rock joints, i.e., there is a certain functional relationship between the shear strength of the rock joints and the absolute value of the relative motion velocity between rock blocks. Therefore, a negative exponential function converging to Q_0 can be used to describe the influence factor of relative motion velocity of the rock joints, and its equation is shown in Eq. (3.10). In Eq. (3.10): Q_0 is the convergence value of the influence factor of relative motion velocity of the rock joints (taking $Q_0 = 0.9$), k is the undetermined coefficient (taking $k = 25$), and $V(t)$ is the relative motion velocity between rock blocks at any moment t of seismic duration; in addition, the physical meanings of other symbols are the same as those above-mentioned.

$$\lambda(t) = Q_0 + (1 - Q_0)e^{-k|V(t)|} \tag{3.10}$$

3.2.2.3 Equation for Solving the SVDC

In conclusion, combined with Eqs. (3.6), (3.9) and (3.10), the equation for solving the SVDC of the rock joints at any moment of seismic duration can be obtained, as shown in Eq. (3.11), and the physical meanings of all symbols in Eq. (3.11) are the same as those above-mentioned. From here we see that the SVDC of the rock joints is a dynamic variable during the entire seismic duration, and its magnitude is determined by the three dynamic response values of the number of cycles, cyclic shear amplitude and relative motion velocity between rock blocks. Additionally, combined with Eqs.

(3.3)–(3.5) and (3.11), the equation of the shear strength characteristic parameters of the rock joints at any moment of seismic duration can be obtained, as shown in Eqs. (3.12)–(3.14), and the physical meanings of all symbols in Eqs. (3.12)–(3.14) are the same as those above-mentioned.

$$\begin{aligned}D_{sc}(t) =&\left[Q_0 + (1 - Q_0)e^{-k|V(t)|}\right]\left\{P_0 + (1 - P_0)e^{-nA(t)}\right.\\&\left.+(1 - P_0)\left[1 - e^{-nA(t)}\right]e^{-mN(t)}\right\}\end{aligned} \tag{3.11}$$

$$\begin{aligned}\varphi(t) = \arctan&\left\{\tan\varphi_0\left\{\left[Q_0 + (1 - Q_0)e^{-k|V(t)|}\right]\left\{P_0 + (1 - P_0)e^{-nA(t)}\right.\right.\right.\\&\left.\left.\left.+(1 - P_0)\left[1 - e^{-nA(t)}\right]e^{-mN(t)}\right\}\right\}\right\}, \alpha = 0\end{aligned} \tag{3.12}$$

$$\begin{aligned}\varphi(t) = \arctan&\left\{\tan(\varphi_0 + \alpha)\left\{\left[Q_0 + (1 - Q_0)e^{-k|V(t)|}\right]\left\{P_0 + (1 - P_0)e^{-nA(t)}\right.\right.\right.\\&\left.\left.\left.+(1 - P_0)\left[1 - e^{-nA(t)}\right]e^{-mN(t)}\right\}\right\}\right\}, \alpha \neq 0\end{aligned} \tag{3.13}$$

$$\begin{aligned}c(t) = c_0&\left[Q_0 + (1 - Q_0)e^{-k|V(t)|}\right]\left\{P_0 + (1 - P_0)e^{-nA(t)}\right.\\&\left.+(1 - P_0)\left[1 - e^{-nA(t)}\right]e^{-mN(t)}\right\}\end{aligned} \tag{3.14}$$

3.3 Analytical Solution of the D_{SC} of the Rock Slope

3.3.1 Equation of the Instantaneous D_{SC}

Considering the strength vibration degradation effect of the rock joints, for the typical rock slope with the penetrating straight rock joints, the derivation process of the equation for solving the instantaneous D_{SC} of slope at any moment of seismic duration will be described in detail below, and the simplified mechanical mode of this kind of slope under the action of seismic load (including horizontal and vertical direction) as shown in Fig. 3.3. Combined with Fig. 3.3, according to the LET and DVM (Zhang & Chen, 2001; Cai et al., 2002; Cheng et al., 2007; Ni et al., 2013; Stianson et al., 2015; Liu et al., 2015; Li et al., 2021), it is defined that the ratio of the projection of the total anti-sliding force vector to the projection of the total sliding force vector of the slip mass in the direction of the sliding surface at any moment of seismic duration is the instantaneous D_{SC} of slope. When the sliding instability failure (ISF) of plane occurs along the sliding surface of slope, the combined external force vector acting on the slip mass is projected to the normal direction of the sliding surface, then the mechanical equilibrium equation is established, and thus the reaction force in the normal direction of the sliding surface at any moment of seismic duration can be solved by this equation, as shown in Eq. (3.15). In Eq. (3.15): at any moment t of seismic duration, $\overrightarrow{R_f}(t)$ is the reaction force vector in the normal direction of the sliding surface, and $\overrightarrow{R_{ef}}(t)$ is the combined external force vector of the slip mass; in addition, $\overrightarrow{m}$ is the normal unit vector outside the sliding surface. Meanwhile, by projecting the combined external force vector acting on the slip mass to the tangential direction of the sliding surface and establishing the mechanical equilibrium equation, the total sliding force vector of the slip mass at any moment of seismic duration can be solved by this equation, as shown in Eq. (3.16). In Eq. (3.16): $\overrightarrow{S_f}(t)$ is the total sliding force vector of the slip mass at any moment t of seismic duration, and $\overrightarrow{n}$ is the tangential unit vector of the sliding surface; in addition, the physical meanings of other symbols are the same as those above-mentioned. Therefore, combined with Eqs. (3.15) and (3.16), when the sliding ISF of plane of slope occurs, the equation (vector form) of the instantaneous D_{SC} of slope at any moment of seismic duration as shown in Eq. (3.17). In Eq. (3.17): $F_s(t)$ is the instantaneous D_{SC} of slope at any moment t of seismic duration, and A is the contact area between the slip mass and sliding surface; in addition, the physical meanings of other symbols are the same as those above-mentioned.

$$\overrightarrow{R_f}(t) = -\left[\overrightarrow{R_{ef}}(t) \cdot \overrightarrow{m}\right] \cdot \overrightarrow{m} \tag{3.15}$$

$$\overrightarrow{S_f}(t) = \left[\overrightarrow{R_{ef}}(t) \cdot \overrightarrow{n}\right] \cdot \overrightarrow{n} \tag{3.16}$$

$$F_s(t) = \left[\left|\overrightarrow{R_f}(t)\right| \tan\varphi(t) + c(t)A\right] \Big/ \left|\overrightarrow{S_f}(t)\right| \tag{3.17}$$

In particular, since this study focuses on the influence of the horizontal seismic load on the stability of the rock slope, i.e., the combined external force of the slip mass is composed of its self-gravity and the horizontal SIF, and in order to the results are more intuitive and clear when programming with the FISH language of the UDEC software in the follow-up study (see Sects. 3.4 and 3.5 for details), Eq. (3.17) is converted to scalar form as shown in Eq. (3.18); furthermore, the vertical SIF is omitted in

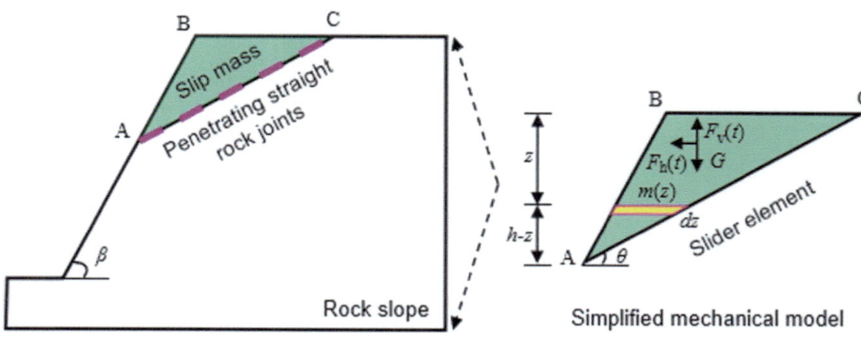

Fig. 3.3 Simplified mechanical model of the typical rock slope with the penetrating straight rock joints subjected to seismic loading (see Sect. 3.3.1 for the specific meanings of the symbols in the this figure)

Eq. (3.18), and thus the equation for solving the instantaneous D_{SC} of the rock slope with the penetrating straight rock joints under the action of horizontal seismic load at any moment of seismic duration can be obtained, as shown in Eq. (3.19). In Eqs. (3.18) and (3.19): at any moment t of seismic duration, $F_v(t)$ is the vertical SIF of the slip mass, $F_h(t)$ is the horizontal SIF of the slip mass, G is the self-gravity of the slip mass, h is the height of the slip mass, θ is the inclination angle of the sliding surface, and $\varphi(t)$ and $c(t)$ are determined by Eqs. (3.12) and (3.14), respectively; in addition, the physical meanings of other symbols are the same as those above-mentioned.

$$F_s(t) = \begin{array}{l}\{\{[G - F_v(t)]\cos\theta - F_h(t)\sin\theta\}\tan\varphi(t) \\ +c(t)h\csc\theta\}\end{array} \Bigg/ \begin{array}{l}\{[G - F_v(t)]\sin\theta \\ +F_h(t)\cos\theta\}\end{array} \tag{3.18}$$

$$F_s(t) = \{[G\cos\theta - F_h(t)\sin\theta]\tan\varphi(t) + c(t)h\csc\theta\}/[G\sin\theta + F_h(t)\cos\theta] \tag{3.19}$$

3.3.2 Equation of the Instantaneous S_{IF}

Based on the DVM (Ni et al., 2013), the horizontal instantaneous SIF of the slip mass can be dynamically obtained by the UDEC software, and the specific solution method is as follows. The combined external force vector of the slip mass at any moment of seismic duration can be obtained by summing the net nodal force vectors of all grid nodes in the slip mass at that moment, and its equation is shown in Eq. (3.20). In Eq. (3.20): at any moment t of seismic duration, $\overrightarrow{F_{udec}}(t)$ is the combined external force vector of the slip mass, n_j is the number of grid nodes of the slip mass, $\overrightarrow{F_{udec}^{g}}(t)$ is the nominal gravity vector of grid nodes of the slip mass, $\overrightarrow{F_{udec}^{l}}(t)$ is the external load vector of the slip mass, $\overrightarrow{F_{udec}^{z}}(t)$ is the internal force vector of the grid element near node of the slip mass, and $\overrightarrow{F_{udec}^{c}}(t)$ is the sub contact force vector only existing between rock blocks of the slip mass. According to the Newton's second law, the equation of the integral acceleration vector of the slip mass at any moment of seismic duration can be obtained, as shown in Eq. (3.21). In Eq. (3.21): $\overrightarrow{a}(t)$ is the integral acceleration vector of the slip mass at any moment t of seismic duration, and m_r is the mass of the slip mass; in addition, the physical meanings of other symbols are the same as those above-mentioned. Additionally, according to the definition of the instantaneous SIF vector, the magnitude of that is equal to the product of the integral acceleration and the mass of the slip mass at any moment of seismic duration, and its direction is opposite to the inertial acceleration direction (Wang et al., 1994). Therefore, the equation of the instantaneous SIF vector can be obtained, as shown in Eq. (3.22). In Eq. (3.22): $\overrightarrow{F_{hv}}(t)$ is the instantaneous SIF vector of slope at any moment t of seismic duration; in addition, the physical meanings of other symbols are the same as those above-mentioned. To sum up, the UDEC software can be applied to obtain the horizontal instantaneous SIF of slope at any moment of seismic duration according to Eq. (3.22), and then it is involved in the limit equilibrium analysis of the dynamic stability of slope; furthermore, the instantaneous D_{SC} of the rock slope subjected to horizontal seismic loading at any moment of seismic duration can be obtained by combining with Eqs. (3.12), (3.14) and (3.19).

$$\overrightarrow{F_{udec}}(t) = \sum_{j=1}^{n_j} \left[\overrightarrow{F_{udec}^{g}}(t) + \overrightarrow{F_{udec}^{l}}(t) + \overrightarrow{F_{udec}^{z}}(t) + \overrightarrow{F_{udec}^{c}}(t) \right] \tag{3.20}$$

$$\vec{a}(t) = \overrightarrow{F_{udec}}(t) \Big/ m_r \tag{3.21}$$

$$\overrightarrow{F_{hv}}(t) = -m_r \vec{a}(t) = -\overrightarrow{F_{udec}}(t) \tag{3.22}$$

3.3.3 Equation of the Final D_{SC}

Based on the above analysis, it can be seen that although the time history curve of the D_{SC} of slope objectively reflects the dynamic change process of the stability of slope with seismic time continuously increasing, it can not directly obtain an effective index for evaluating the dynamic stability of slope. Relevant scientific study (Liu et al., 2003) shows that since the element shear strain is in a continuous updating state during the dynamic response analysis of a large step (it can not be compared with the shear strain initially assumed in the calculation), the "stride iteration method" is generally used to realize this dynamic response analysis. Therefore, it is often necessary to assume an equivalent and constant shear strain to replace the change process, and several scholars have proposed a conversion equation, as shown in Eq. (3.23); in Eq. (3.23): γ_{eq} is the equivalent and constant hypothetical shear strain, ϑ is the conversion coefficient (taking $\vartheta = 0.65$), and $\gamma_{\max}$ is the maximum value of the element shear strain in the seismic duration. The conversion coefficient can be determined from the perspective of probability analysis or energy change according to the seismic load characteristics (usually empirical), and most scholars have proposed different conversion coefficient values, but the values are between 0.55 ~ 0.75. Meanwhile, this study also shows that the D_{SC} of slope will continue to fluctuate up and down with the seismic excitation process, and the minimum dynamic stability coefficient ($DSC_{\min}$) will appear in this process. However, when this coefficient or the static stability coefficient (S_{SC}) is taken as the final stability coefficient of slope, the corresponding sliding surface may not be the same (caused by the continuous change of dynamic shear strain).

$$\gamma_{eq} = \vartheta \gamma_{\max} \tag{3.23}$$

In view of this, considering that the slope may not be completely destabilized under the instantaneous impact action of seismic load, therefore, 0.65 times of the difference between the S_{SC} and the DSC_{min} is taken as the average amplitude of the final stability coefficient, so as to reflect the deviation amplitude of the D_{SC} due to seismic excitation more objectively. That is, the minimum average safety coefficient method ($ASCM_{min}$) is used to solve the value of the final dynamic stability evaluation index of slope, as shown in Eq. (3.24); in Eq. (3.24): F_{dei} is the evaluation index of the dynamic stability of slope, F_{s0} is the S_{SC} of slope, and $F_{d\min}$ is the minimum stability coefficient in the time history curve of the dynamic stability of slope. Especially, this study will refer to this method to solve the value of the final dynamic stability evaluation index of the rock slope subjected to seismic loading.

$$F_{dei} = F_{s0} - 0.65(F_{s0} - F_{d\min}) \tag{3.24}$$

3.4 Numerical Realization of the Dynamic Stability Algorithm

3.4.1 Calculation Scheme

Firstly, taking the dynamic stability analysis of the typical rock slope with the penetrating straight rock joints subjected to horizontal seismic loading as an example, and using the UDEC software to establish the discrete element numerical calculation model; thus, the necessity of considering the strength vibration degradation effect of the rock joints in the process of seismic load can be revealed through the example analysis. Then, selecting the dynamic stability analysis of the rock slope with the penetrating straight rock joints under the background of Wenchuan seismic wave (horizontal direction) for the case study, and thus the feasibility of the algorithm can be verified by comparing and analyzing the numerical calculation and shaking table model test results. Furthermore, fully considering the changes of the strength characteristic parameters of the rock joints (initial cohesion and internal friction angle), geometric parameters of slope (slope angle and dip angle of bedding plane) and dynamic characteristic parameters (seismic load amplitude and seismic load frequency), so as to explore the influence of different factors on the dynamic stability response rules of slope.

Based on this, and considering the influence of the boundary effect on slope, the specific geometric dimensions, mesh division and boundary conditions of the discrete element numerical calculation model are shown in Fig. 3.4. Among them, the rock mass of the model slope was divided into several triangular mesh elements, and their geometric dimensions are all 10 m. Both in the initial equilibrium and dynamic analysis stages, the top (surface) of the model slope was set as the free boundary; in the initial equilibrium stage, the left and right sides were set as the horizontal fixed boundary, and the bottom was set as the vertical fixed boundary; in the dynamic analysis stage, the left and right sides were set as the free field boundary (semi-infinite condition), and the horizontal direction (loading direction) and vertical direction of bottom were set as the viscous boundary (reducing reflection superposition) and fixed boundary, respectively; in the initial equilibrium and dynamic analysis stages, the linear elastic model and Mohr–Coulomb model were adopted for the rock mass of the model slope, respectively; both in the initial equilibrium and dynamic analysis stages, the Contact-Coulomb slip model and linear elastic model were adopted for the rock joints and free field (viscous) boundary, respectively.

In particular, to investigate the dynamic stability of the example analysis slope, the simple harmonic wave was used to simulate the seismic load, i.e., the sinusoidal shear wave with acceleration time history function of $a = \lambda' \sin(2\pi f' t')$ was used to apply the seismic load. Wherein, the amplitude λ' is 0.10 g ($g = 10$ m/s^2, similarly hereinafter), the frequency f' is 5 Hz, and the duration t' is 15 s. According to Sect. 3.2.2, the values of relevant parameters in the equation (as shown in Eq. (3.11)) of the SVDC of the rock joints in the numerical calculation are as follows: the convergence values $P_0 = 0.49386$ and $Q_0 = 0.9$; the undetermined coefficients $m = 0.00146$, $n = 0.719$ and $k = 25$. Additionally, the physical–mechanical parameters of the rock mass (boundary) and the rock joints are shown in Tables 3.1 and 3.2,

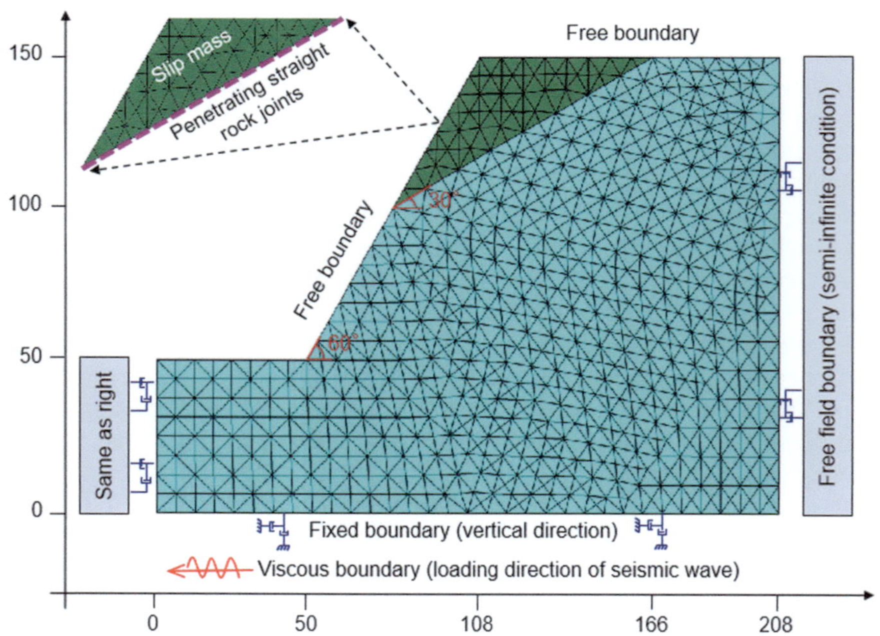

Fig. 3.4 Discrete element numerical calculation model established by UDEC for the example analysis slope

Table 3.1 Physical–mechanical parameters of the rock mass and boundary

Item	Density (g/cm^3)	Bulk modulus (GPa)	Shear modulus (GPa)	Cohesion (kPa)	Internal friction angle (°)	Tensile strength (kPa)
Rock mass	2.65	4.23	2.66	5260	44.53	526
Boundary	2.65	4.23	2.66	-	-	-

Table 3.2 Physical–mechanical parameters of the rock joints

Item	Normal stiffness (GPa/m)	Tangential stiffness (GPa/m)	Cohesion (kPa)	Internal friction angle (°)	Tensile strength (kPa)
Rock joints	16	12	90	30	60

respectively; the specific working conditions of the analysis of factors affecting the dynamic stability of the rock slope are shown in Tables 3.3, 3.4 and 3.5 in Sect. 3.5 for details.

3.4.2 Calculation Procedure

The UDEC software was used as the calculation engine, and its built-in FISH language was used to program and realize the dynamic algorithm of the D_{SC} of slope considering the strength vibration degradation effect of the rock joints. The specific flow chart of numerical implementation is shown in Fig. 3.5.

3.4.3 Response Characteristics of the Dynamic Stability

For the example analysis slope, the time history curves of the $SVDC_{min}$ of the rock joints and the D_{SC} of slope during the seismic action are shown in Figs. 3.6 and 3.7, respectively. It can be seen from Fig. 3.6 that the $SVDC_{min}$ of the rock joints gradually decays in the form of negative exponential function with the seismic excitation time, and the fitting effect of the function expression is good, which indicates that there is an approximate unique relationship between the seismic excitation time and the $SVDC_{min}$ in a certain interval. In particular, in the early stage of seismic action, the $SVDC_{min}$ is mainly affected by the vibration wear between rock blocks, resulting in a relatively rapid attenuation rate; while in the later stage of seismic action, the $SVDC_{min}$ does not continue to significantly attenuate, and it only fluctuates up and down (tends to be stable) with the influence of the relative motion velocity between rock blocks.

Table 3.3 Statistics of calculation results of the dynamic stability of slope under different strength characteristic parameters

Initial cohesion (kPa)	Initial internal friction angle (°)	Value of the final dynamic stability evaluation index	Nephogram of the maximum shear strain increment
60	30	1.32–0.65 × (1.32–0.9025) = 1.0486	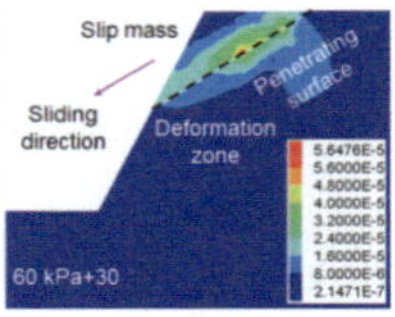
75	30	1.39–0.65 × (1.39–0.9122) = 1.0794	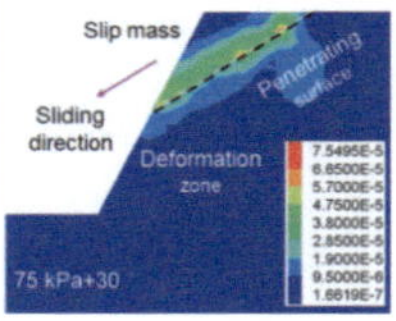
90	30	1.47–0.65 × (1.47–0.9132) = 1.1081	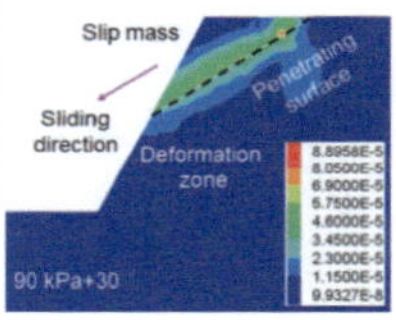
105	30	1.55–0.65 × (1.55–0.9167) = 1.1383	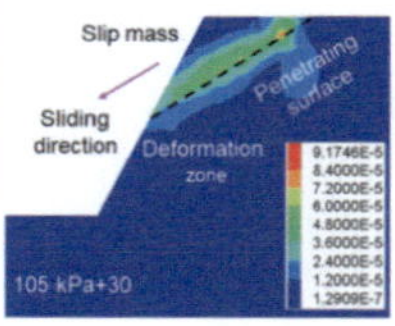
120	30	1.63–0.65 × (1.63–0.9194) = 1.1681	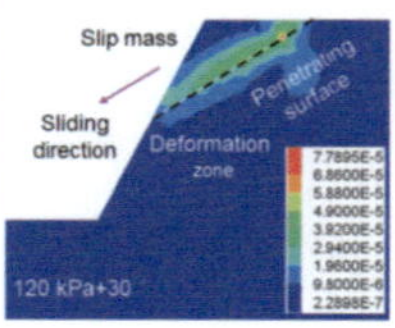
90	20	1.10–0.65 × (1.10–0.8877) = 0.9620	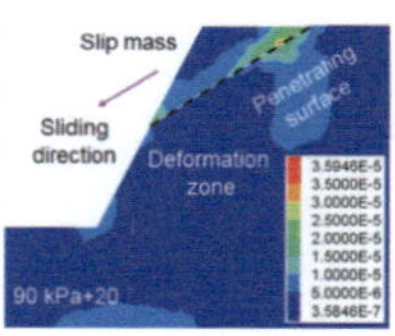

(continued)

Table 3.3 (continued)

Initial cohesion (kPa)	Initial internal friction angle (°)	Value of the final dynamic stability evaluation index	Nephogram of the maximum shear strain increment
90	25	1.27–0.65 × (1.27–0.8974) = 1.0278	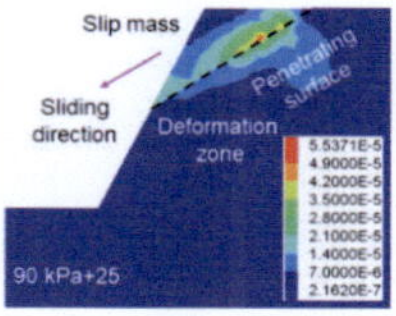
90	35	1.69–0.65 × (1.69–0.9142) = 1.1858	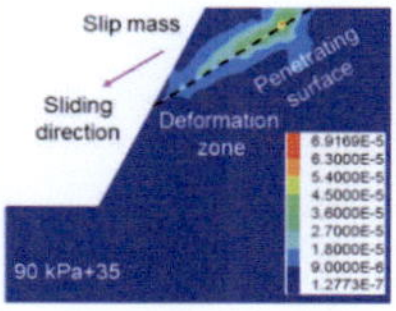
90	40	1.93–0.65 × (1.93–0.9248) = 1.2766	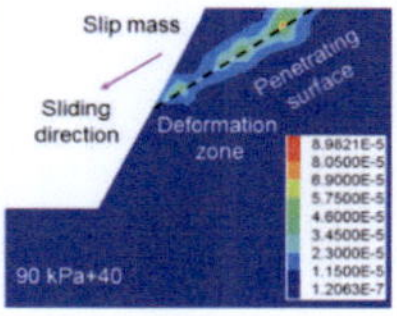

It can be seen from Fig. 3.7 that for the example analysis slope, the fluctuation range of the D_{SC} of slope is 1.0118 ~ 5.1796 without considering the strength vibration degradation effect of the rock joints, and the D_{SC} is maintained at about 1.4693 after the termination of seismic excitation. When considering the strength vibration degradation effect of the rock joints, the shear strength characteristic parameters of the rock joints gradually attenuate dynamically during the seismic duration, resulting in the D_{SC} of slope continuously decreasing in the dynamic fluctuation process (the fluctuation range is 0.9132 ~ 4.7960). Meanwhile, the D_{SC} also decreases after the termination of seismic excitation (maintained at about 1.4670), i.e., an seismic lasting 15 s has caused irreversible permanent cumulative damage to the stability of the rock slope. From here we see that the calculation results obtained by considering the strength vibration degradation effect of the rock joints in the dynamic stability analysis of slope are more in line with the general natural rules.

Furthermore, according to Fig. 3.7, the $ASCM_{min}$ can be used to solve the value of the final dynamic stability evaluation index of slope (as shown in Eq. (3.24), and the S_{SC} of slope is 1.47). Based on this, it can be obtained that the values of the final dynamic stability evaluation index of the example analysis slope are 1.1722 and 1.1081, respectively when the strength vibration degradation effect of the rock joints is not considered and considered. Although the difference between the two is not large, it can still reveal a certain change trend, i.e., the value of the final

Table 3.4 Statistics of calculation results of the dynamic stability of slope under different geometric parameters of slope

Slope angle (°)	Dip angle of bedding plane (°)	Value of the final dynamic stability evaluation index	Nephogram of the maximum shear strain increment
50	30	1.74–0.65 × (1.74–0.9385) = 1.2190	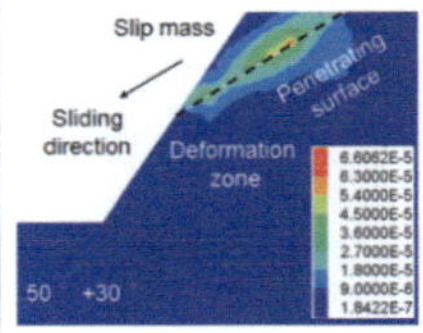
55	30	1.59–0.65 × (1.59–0.9251) = 1.1578	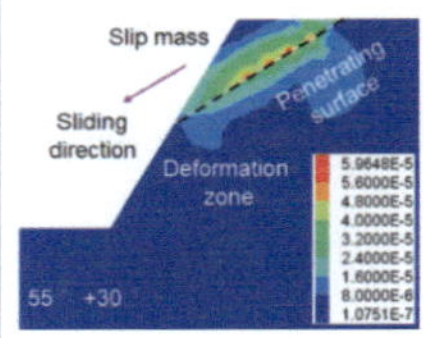
60	30	1.47–0.65 × (1.47–0.9132) = 1.1081	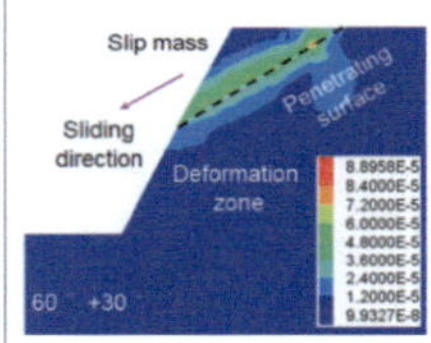
65	30	1.36–0.65 × (1.36–0.8987) = 1.0601	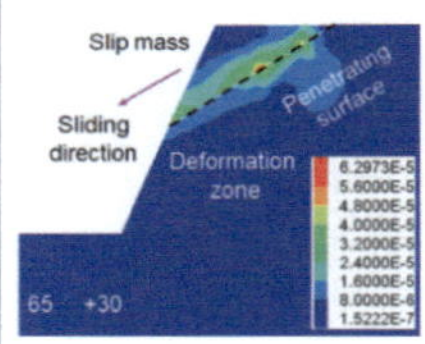
70	30	1.22–0.65 × (1.22–0.8799) = 0.9814	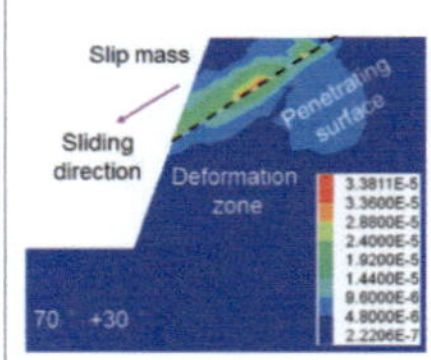
60	28	1.59–0.65 × (1.59–0.9194) = 1.1541	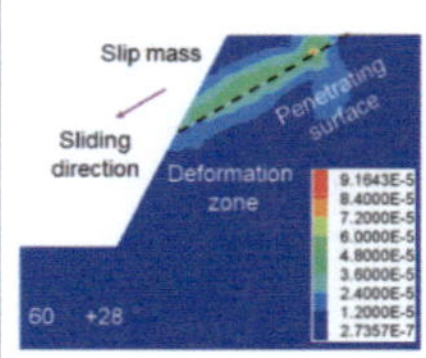

(continued)

Table 3.4 (continued)

Slope angle (°)	Dip angle of bedding plane (°)	Value of the final dynamic stability evaluation index	Nephogram of the maximum shear strain increment
60	32	1.37–0.65 × (1.37–0.9050) = 1.0678	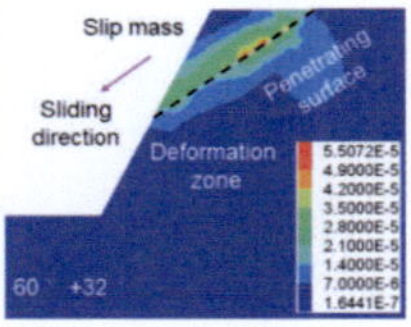
60	34	1.23–0.65 × (1.23–0.8999) = 1.0154	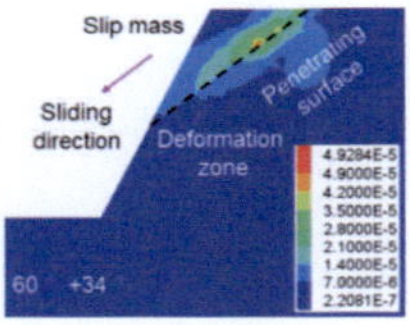
60	36	1.17–0.65 × (1.17–0.8603) = 0.9687	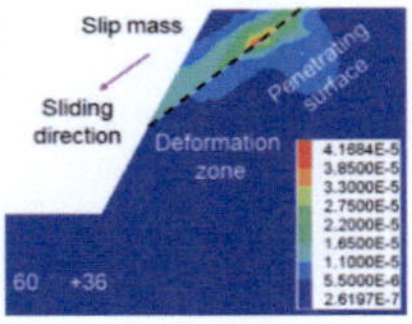

dynamic stability evaluation index of slope calculated when the strength vibration degradation effect of the rock joints is considered during the seismic duration is smaller than that when the strength vibration degradation effect is not considered, and this result accords with the actual rules.

3.4.4 Example Verification

Combined with the relevant scientific study (Gine, 2011; Yang et al., 2021), to verify the feasibility of the dynamic algorithm for solving the D_{SC} of slope considering the strength vibration degradation effect of the rock joints proposed in this study, the horizontal wave (natural wave) of Wenchuan seismic recorded by an seismic monitoring station was selected as the seismic load to conduct a case study on the dynamic stability of the rock slope. In particular, the curve of the acceleration time history of seismic load applied to the example verification slope after high-frequency filtering and baseline correction of the natural wave is shown in Fig. 3.8a (intercepted 0 ~ 50 s, and the strong seismic excitation time is about 20 s), and the corresponding curve of the Fourier amplitude spectrum is shown in Fig. 3.8b (intercepted 0 ~ 100 Hz, and the

Table 3.5 Statistics of calculation results of the dynamic stability of slope under different dynamic characteristic parameters; and $g = 10$ m/s^2

Seismic load amplitude ($\times 10$ m/s^2)	Seismic load frequency (Hz)	Value of the final dynamic stability evaluation index	Nephogram of the maximum shear strain increment
0.02	5	1.47–0.65 × (1.47–1.2097) = 1.3008	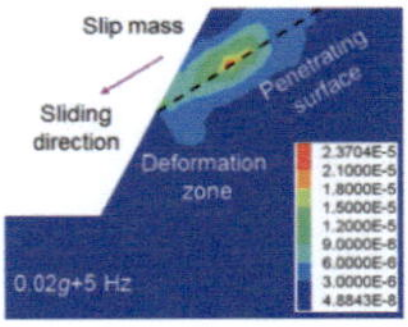
0.04	5	1.47–0.65 × (1.47–1.0228) = 1.1793	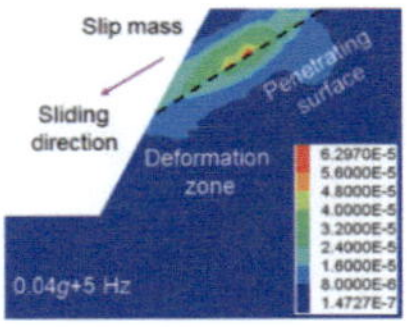
0.08	5	1.47–0.65 × (1.47–0.9146) = 1.1090	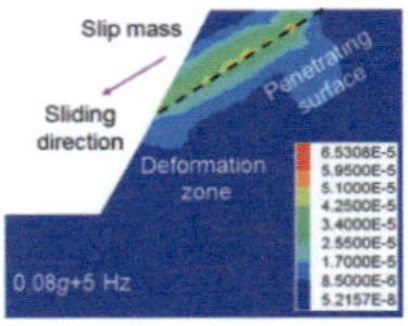
0.10	5	1.47–0.65 × (1.47–0.9132) = 1.1081	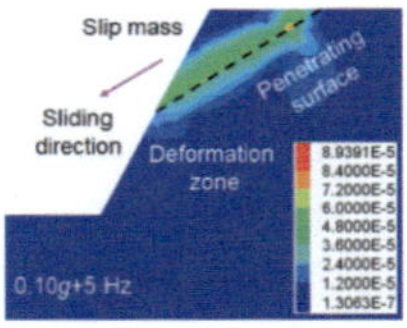
0.12	5	1.47–0.65 × (1.47–0.9109) = 1.1066	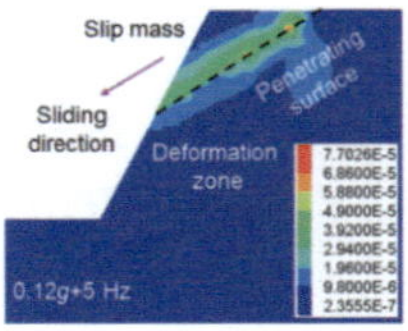
0.10	3	1.47–0.65 × (1.47–0.9199) = 1.1124	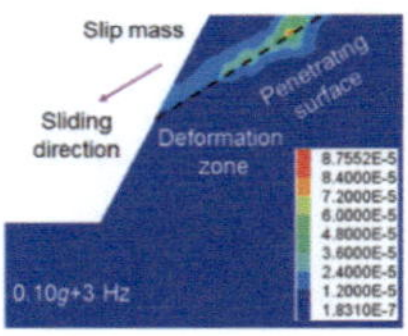

(continued)

Table 3.5 (continued)

Seismic load amplitude ($\times 10$ m/s^2)	Seismic load frequency (Hz)	Value of the final dynamic stability evaluation index	Nephogram of the maximum shear strain increment
0.10	7	1.47–0.65 × (1.47–0.9122) = 1.1074	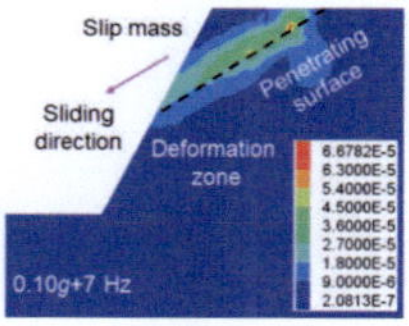
0.10	8	1.47–0.65 × (1.47–0.9111) = 1.1067	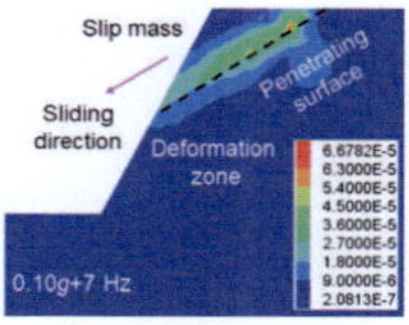
0.10	10	1.47–0.65 × (1.47–0.9097) = 1.1058	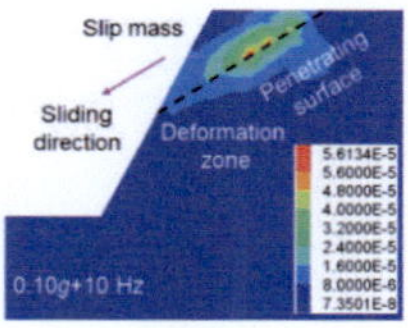

main frequency range is 7 ~ 12 Hz); meanwhile, the relevant physical–mechanical parameters and dynamic characteristic parameters in the numerical calculation are consistent with those described in study (Gine, 2011; Yang et al., 2021).

Based on this, the time history curve of the D_{SC} of slope under the action of horizontal Wenchuan wave can be obtained by applying the algorithm proposed in this study, as shown in Figs. 3.9 and 3.10 is the comparison between the numerical calculation results of the DCD and ISF modes of slope during the seismic duration (including the nephograms of the maximum shear strain increment of slope and the overall displacement of the slip mass) and the shaking table test observations.

It can be seen from Figs. 3.9 and 3.10 that during the seismic duration, the D_{SC} of slope gradually increases with time and fluctuates up and down with the fluctuation range of 0.4592 ~ 2.8191, and the fluctuation phenomenon of the coefficient is very significant during the period of the strong seismic excitation. Meanwhile, the S_{SC} of slope is 1.25, and according to the calculation with the $ASCM_{min}$ (as shown in Eq. (3.24)), the value of the final dynamic stability evaluation index of slope is 0.7360, i.e., the calculation results show that the slope has been completely destabilized under the action of horizontal Wenchuan wave. Additionally, the evolution process of the DCD and ISF of the accumulation mass in the shaking table model test can be summarized as follows: a small amount of rock and soil particles on the slope are firstly exfoliated; then, the settlement of the top of the accumulation mass gradually increases and its lateral damage cracks continue to extend and widen;

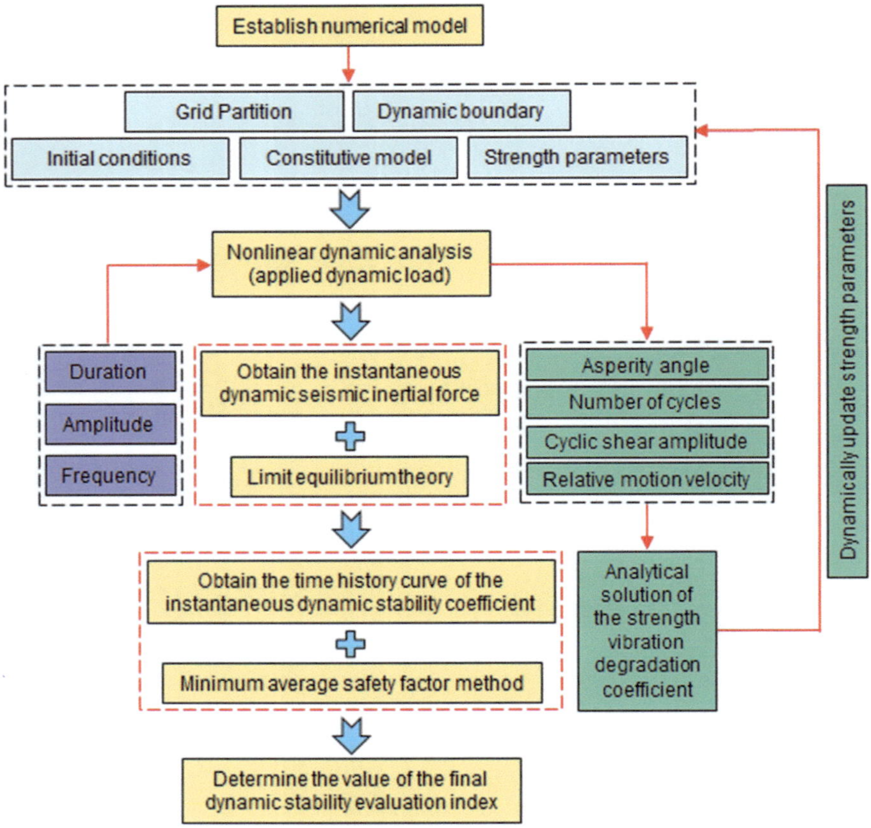

Fig. 3.5 Flow chart of numerical implementation

subsequently, the number of damage cracks on the top of the accumulation mass suddenly increases and the overall sliding ISF occurs along the penetrating straight rock joints (the residual marks on the sliding surface are extremely significant). In conclusion, the strength vibration degradation effect of the rock joints has a great impact on the dynamic stability of slope. Especially, the necessity of considering the strength vibration degradation effect of the rock joints in the dynamic stability analysis of slope and the feasibility of the algorithm derived in above for solving the D_{SC} of slope considering the strength vibration degradation effect of the rock joints have been well verified.

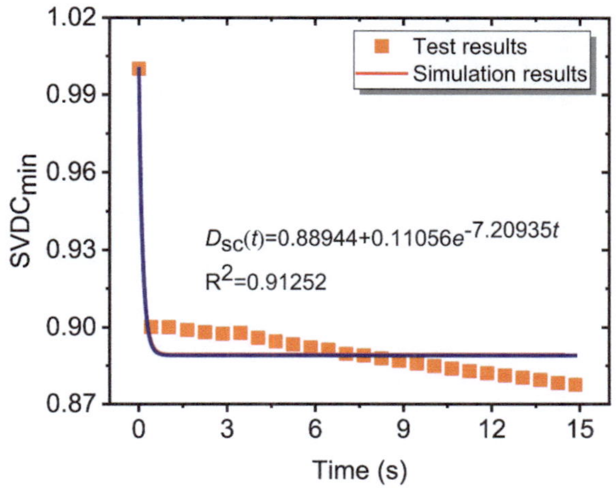

Fig. 3.6 Tme history curve of the $SVDC_{min}$ of the rock joints for the example analysis slope: $SVDC_{min}$ represents the minimum SVDC (see Sect. 3.2.1 for the specific meanings of the symbols in this figure)

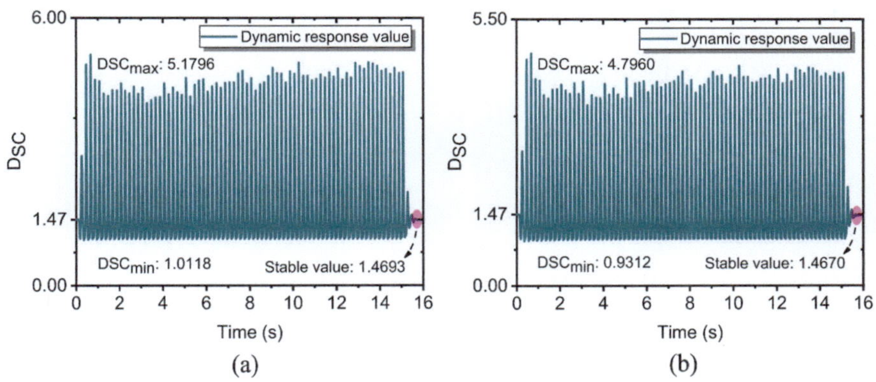

Fig. 3.7 Time history curve of the D_{SC} for the example analysis slope: **a** Strength vibration degradation effect is not considered, and **b** Strength vibration degradation effect is considered; DSC_{max} and DSC_{min} represent the maximum and minimum D_{SC}, respectively

3.5 Analysis of Influence Parameters on the Dynamic Stability

3.5.1 Strength Characteristic Parameters

This part mainly explores the influence of the strength characteristic parameters (initial cohesion and internal friction angle) of the rock joints on the dynamic stability response of slope when considering the strength vibration degradation effect of

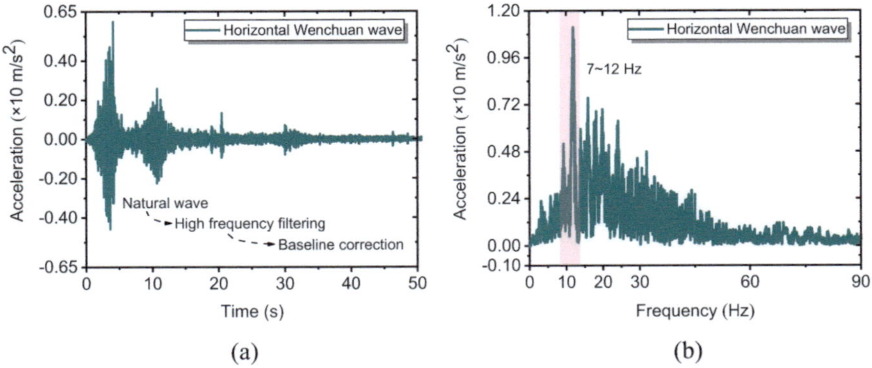

Fig. 3.8 Curves of the acceleration time history and fourier amplitude spectrum subjected to seismic loading for the example verification slope: **a** acceleration time history, and **b** fourier amplitude spectrum

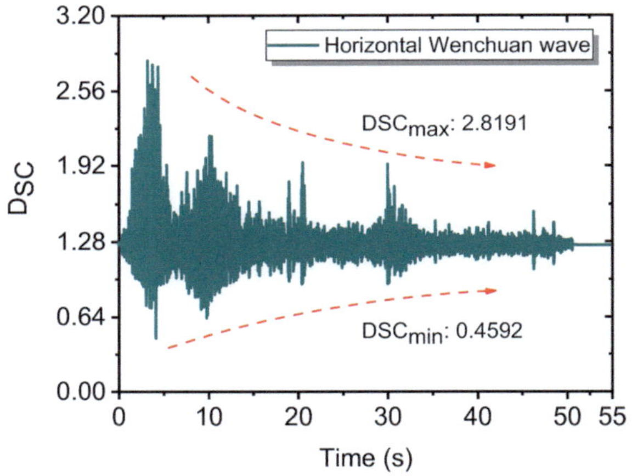

Fig. 3.9 Time history curve of the D_{SC} for the example verification slope considering the strength vibration degradation effect of the rock joints: DSC_{max} and DSC_{min} represent the maximum and minimum D_{SC}, respectively

the rock joints. Among them, the values of the initial cohesion include 60 kPa, 75 kPa, 90 kPa, 105 kPa and 120 kPa, the values of the initial internal friction angle include 20°, 25°, 30°, 35° and 40°, and the relevant physical–mechanical parameters and dynamic characteristic parameters in the numerical calculation can be seen in Sect. 3.4.1 for details. Additionally, the time history curves of the $SVDC_{min}$ of the rock joints and the DSC_{min} of slope, and the change curves of the D_{SC} of slope under different strength characteristic parameters are shown in Figs. 3.11, 3.12 and 3.13, respectively; the relevant statistics of calculation results of the dynamic stability of slope are shown in Table 3.3.

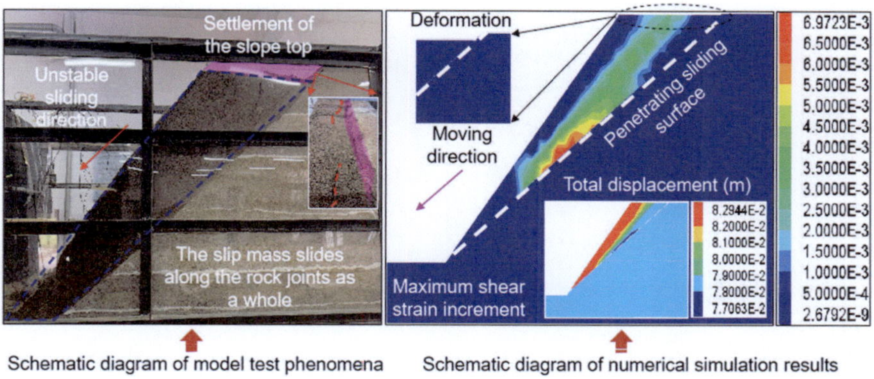

Fig. 3.10 Comparison of the DCD and ISF modes for the example verification slope

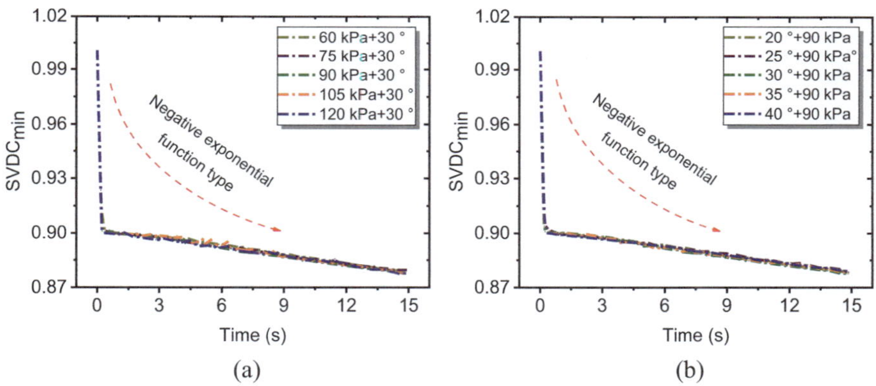

Fig. 3.11 Time history curve of the $SVDC_{min}$ of the rock joints under different strength characteristic parameters: **a** initial cohesion, and **b** initial internal friction angle; $SVDC_{min}$ represents the minimum SVDC

It can be seen from Fig. 3.11 that the $SVDC_{min}$ of the rock joints gradually attenuates in the form of negative exponential function with the increase of the seismic action time under different initial cohesions or internal friction angles. Meanwhile, during the seismic duration, the $SVDC_{min}$ of the rock joints decreases rapidly in the early stage mainly due to the influence of the vibration wear between rock blocks, and tends to be stable in the later stage mainly due to the influence of the relative motion velocity between rock blocks. In general, the $SVDC_{min}$ of the rock joints increases with the increase of the initial cohesion or internal friction angle during the seismic duration, i.e., the degradation degree of the shear strength of the rock joints weakens, which is conducive to improving the overall stability of slope.

It can be seen from Figs. 3.12 and 3.13 that the DSC_{min} of slope shows a trend of first steep drop and then basically stable with the continuous increase of the seismic action time under different initial cohesions or internal friction angles. Meanwhile,

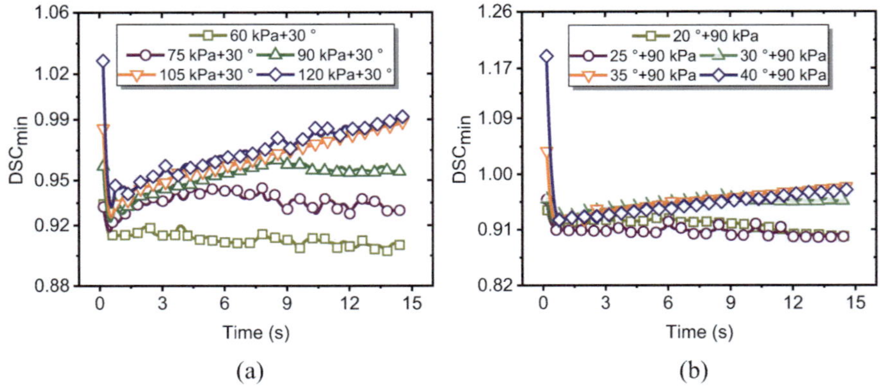

Fig. 3.12 Time history curve of the DSC_{min} of slope under different strength characteristic parameters: **a** initial cohesion, and **b** initial internal friction angle; DSC_{min} represents the minimum D_{SC}

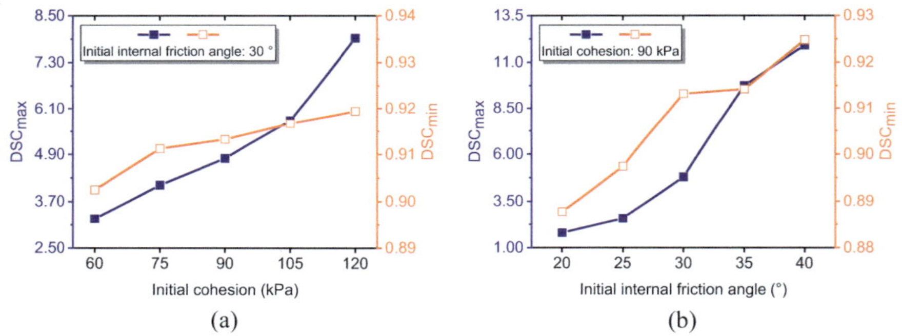

Fig. 3.13 Curve of the D_{SC} of slope under different strength characteristic parameters: **a** initial cohesion, and **b** initial internal friction angle; DSC_{max} and DSC_{min} represent the maximum and minimum D_{SC}, respectively

the influence of the change for the initial cohesion of the rock joints on the fluctuation of the time history curve during the seismic duration is more obvious than the initial internal friction angle, and the DSC_{min} of slope increases with the increase of the initial cohesion or internal friction angle as a whole. Additionally, the D_{SC} of slope shows a fluctuating trend with the increase of the seismic action time under different initial cohesions or internal friction angles, and the fluctuation degree is significant with the increase of the initial cohesion or internal friction angle. Especially, when the initial cohesion is 60 kPa, 75 kPa, 90 kPa, 105 kPa and 120 kPa, respectively, the fluctuation range is 0.9025 ~ 3.2573, 0.9122 ~ 4.1186, 0.9132 ~ 4.7960, 0.9167 ~ 5.7835 and 0.9194 ~ 7.9240; when the initial internal friction angle is 20°, 25°, 30°, 35° and 40°, respectively, the fluctuation range is 0.8877 ~ 1.7987, 0.8974 ~ 2.5712, 0.9132 ~ 4.7960, 0.9142 ~ 9.7541 and 0.9248 ~ 11.9158.

It can be seen from Table 3.3 that according to the $ASCM_{min}$ (as shown in Eq. (3.24)), when the initial cohesion is 60 kPa, 75 kPa, 90 kPa, 105 kPa and 120 kPa, respectively (at this time, the S_{SC} of slope is 1.32, 1.39, 1.47, 1.55 and 1.63, respectively), the values of the final dynamic stability evaluation index of slope are 1.0486, 1.0794, 1.1081, 1.1383 and 1.1681, respectively. Meanwhile, when the initial internal friction angle is 20°, 25°, 30°, 35° and 40°, respectively (at this time, the S_{SC} of slope is 1.10, 1.27, 1.47, 1.69 and 1.93, respectively), the values of the final dynamic stability evaluation index of slope are 0.9685, 1.0278, 1.1081, 1.1858 and 1.2766, respectively. Additionally, the final dynamic stability of slope increases with the increase of the initial cohesion or internal friction angle, and the variation characteristics of the nephogram of the maximum shear strain increment of slope under different strength characteristic parameters have obvious differences and reasonable regularity.

3.5.2 Geometric Parameters of Slope

This part mainly explores the influence of the geometric parameters of slope (slope angle and dip angle of bedding plane) of the rock joints on the dynamic stability response of slope when considering the strength vibration degradation effect of the rock joints. Among them, the values of the slope angle include 50°, 55°, 60°, 65° and 70°, the values of the dip angle of bedding plane include 28°, 30°, 32°, 34° and 36°, and the relevant physical–mechanical parameters and dynamic characteristic parameters in the numerical calculation can be seen in Sect. 3.4.1 for details. Additionally, the time history curves of the $SVDC_{min}$ of the rock joints and the DSC_{min} of slope, and the change curves of the D_{SC} of slope under different geometric parameters of slope are shown in Figs. 3.14, 3.15 and 3.16, respectively; the relevant statistics of calculation results of the dynamic stability of slope are shown in Table 3.4.

It can be seen from Fig. 3.14 that the $SVDC_{min}$ of the rock joints gradually attenuates in the form of negative exponential function with the increase of the seismic action time under different slope angles or dip angles of bedding plane. Meanwhile, the $SVDC_{min}$ of the rock joints is obviously affected by the vibration wear and relative motion velocity of rock blocks during the seismic duration. In general, the $SVDC_{min}$ of the rock joints decreases with the increase of the slope angle or dip angle of bedding plane during the seismic duration, i.e., the degradation degree of the shear strength of the rock joints aggravates, which has a negative impact on the overall stability of slope.

It can be seen from Figs. 3.15 and 3.16 that the DSC_{min} of slope shows a trend of first steep drop and then basically stable (it shows the development characteristics of slightly slow rise at the same time) with the continuous increase of the seismic action time under different slope angles or dip angles of bedding plane. Meanwhile, the influence of the change for the dip angle of bedding plane on the fluctuation of the time history curve during the seismic duration is more obvious than the slope angle, and the DSC_{min} of slope decreases with the increase of the slope angle or dip

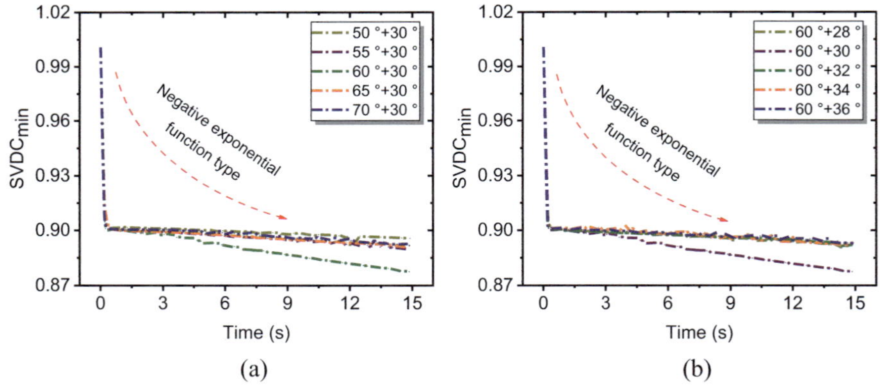

Fig. 3.14 Time history curve of the $SVDC_{min}$ of the rock joints under different geometric parameters of slope: **a** slope angle, and **b** dip angle of bedding plane; $SVDC_{min}$ represents the minimum SVDC

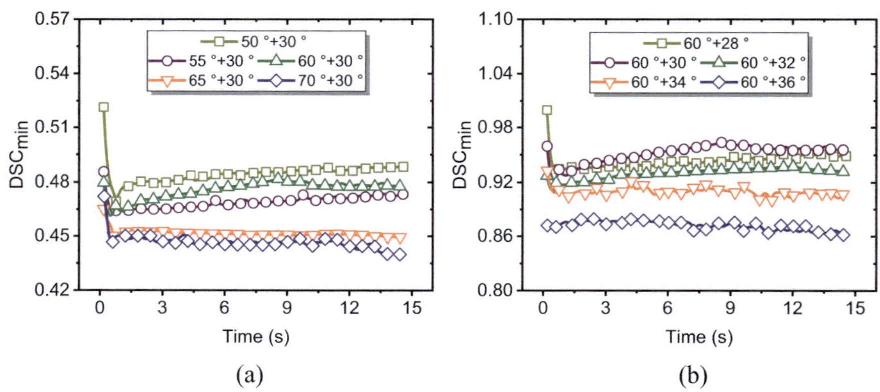

Fig. 3.15 Time history curve of the DSC_{min} of slope under different geometric parameters of slope: **a** slope angle, and **b** dip angle of bedding plane; DSC_{min} represents the minimum D_{SC}

angle of bedding plane as a whole. Additionally, the D_{SC} of slope shows a fluctuating trend with the increase of the seismic action time under different slope angles or dip angles of bedding plane, and the fluctuation degree weakens with the increase of the slope angle or dip angle of bedding plane. Especially, when the slope angle is 50°, 55°, 60°, 65° and 70°, respectively, the fluctuation range is 0.9385 ~ 11.6061, 0.9251 ~ 6.2446, 0.9132 ~ 4.7960, 0.8987 ~ 3.7477 and 0.8799 ~ 2.3531; when the dip angle of bedding plane is 28°, 30°, 32°, 34° and 36°, respectively, the fluctuation range is 0.9194 ~ 6.1893, 0.9132 ~ 4.7960, 0.9050 ~ 3.3810, 0.8999 ~ 2.2031 and 0.8603 ~ 1.9621.

It can be seen from Table 3.4 that according to the $ASCM_{min}$ (as shown in Eq. (3.24)), when the slope angle is 50°, 55°, 60°, 65° and 70°, respectively (at this time, the S_{SC} of slope is 1.74, 1.59, 1.47, 1.36 and 1.22, respectively), the values

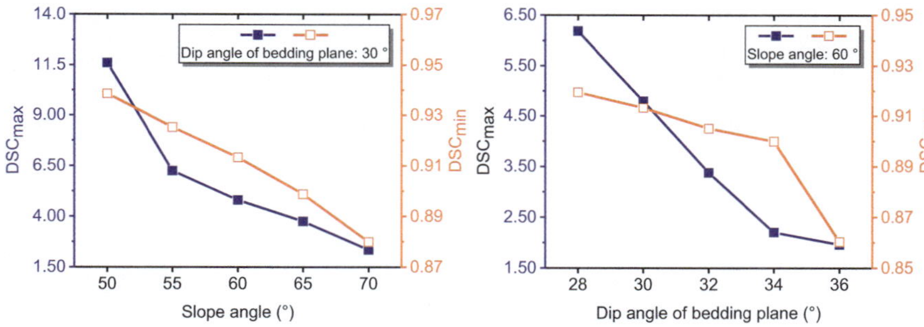

(a) (b)

Fig. 3.16 Curve of the D_{SC} of slope under different geometric parameters of slope: **a** slope angle, and **b** dip angle of bedding plane; DSC_{max} and DSC_{min} represent the maximum and minimum D_{SC}, respectively

of the final dynamic stability evaluation index of slope are 1.2190, 1.1578, 1.1081, 1.0601 and 0.9814, respectively. Meanwhile, when the dip angle of bedding plane is 28°, 30°, 32°, 34° and 36°, respectively (at this time, the S_{SC} of slope is 1.59, 1.47, 1.37, 1.23 and 1.17, respectively), the values of the final dynamic stability evaluation index of slope are 1.1541, 1.1081, 1.0678, 1.0154 and 0.9687, respectively. Additionally, the final dynamic stability of slope decreases with the increase of slope angle or dip angle of bedding plane, and the variation characteristics of the nephogram of the maximum shear strain increment of slope under different geometric parameters of slope have obvious differences and reasonable regularity.

3.5.3 Dynamic Characteristic Parameters

This part mainly explores the influence of the dynamic characteristic parameters (seismic load amplitude and seismic load frequency) of the rock joints on the dynamic stability response of slope when considering the strength vibration degradation effect of the rock joints. Among them, the values of the seismic load amplitude include 0.02 *g*, 0.04 *g*, 0.08 *g*, 0.10 *g* and 0.12 *g*, the values of the seismic load frequency include 3 Hz, 5 Hz, 7 Hz, 8 Hz and 10 Hz, and the relevant physical–mechanical parameters and dynamic characteristic parameters in the numerical calculation can be seen in Sect. 3.4.1 for details. Additionally, the time history curves of the $SVDC_{min}$ of the rock joints and the DSC_{min} of slope, and the change curves of the D_{SC} of slope under different geometric parameters of slope are shown in Figs. 3.17, 3.18 and 3.19, respectively; the relevant statistics of calculation results of the dynamic stability of slope are shown in Table 3.5.

It can be seen from Fig. 3.17 that the $SVDC_{min}$ of the rock joints gradually attenuates in the form of negative exponential function with the increase of the seismic action time under different seismic load amplitudes or seismic load frequencies.

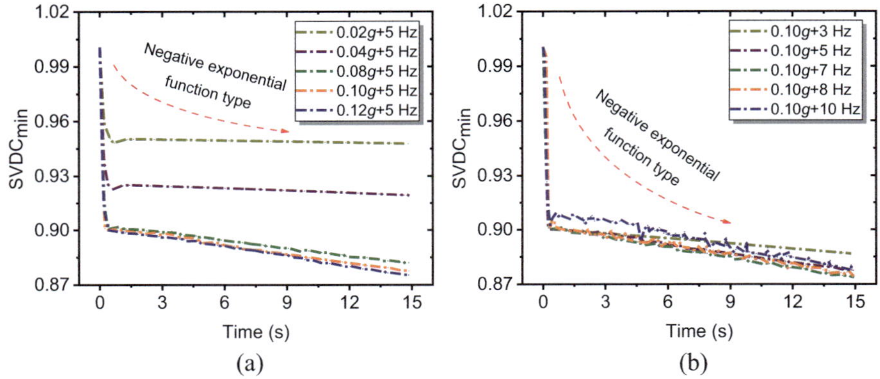

Fig. 3.17 Time history curve of the $SVDC_{min}$ of the rock joints under different dynamic characteristic parameters: **a** seismic load amplitude, and **b** seismic load frequency; $SVDC_{min}$ represents the minimum SVDC; and $g = 10\ m/s^2$

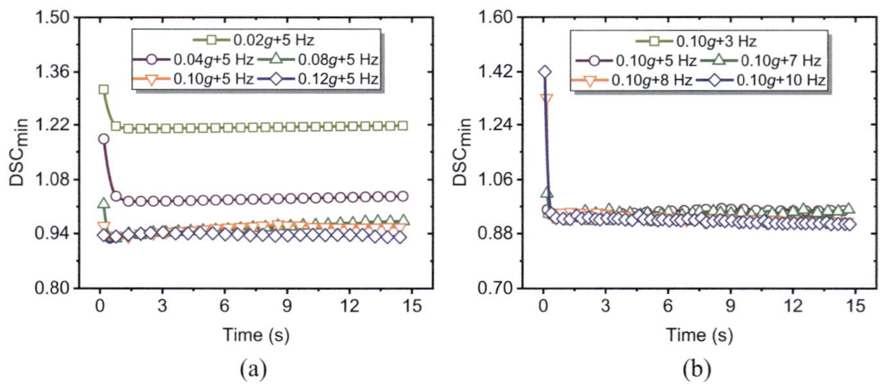

Fig. 3.18 Time history curve of the DSC_{min} of slope under different dynamic characteristic parameters: **a** seismic load amplitude, and **b** seismic load frequency; DSC_{min} represents the minimum D_{SC}; and $g = 10\ m/s^2$

Meanwhile, during the seismic duration, the vibration wear phenomenon between rock blocks caused by the change of the seismic load amplitude is obvious, and the relative motion velocity between rock blocks is also intense. In general, the $SVDC_{min}$ of the rock joints decreases with the increase of the seismic load amplitude or seismic load frequency during the seismic duration; the fluctuation of the attenuation curve for the $SVDC_{min}$ of the rock joints caused by the change of the seismic load frequency is more obvious than that of the seismic load amplitude, and the difference is weak, i.e., the deterioration degree of the shear strength of the rock joints is aggravated, which has a negative impact on the overall stability of slope.

It can be seen from Figs. 3.18 and 3.19 that the DSC_{min} of slope shows a trend of first steep drop and then basically stable with the continuous increase of the seismic

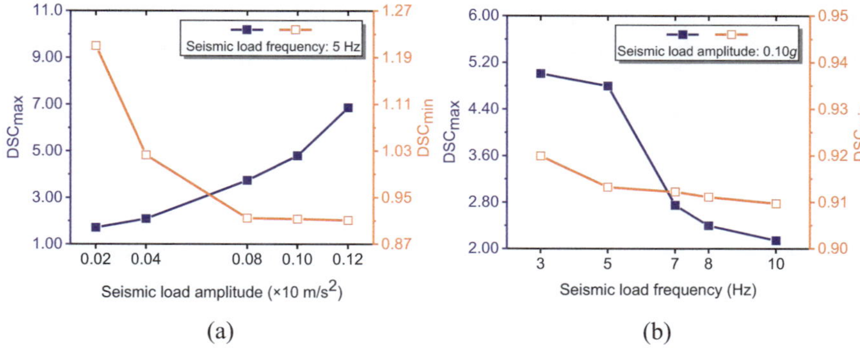

Fig. 3.19 Curve of the D_{SC} of slope under different dynamic characteristic parameters: **a** seismic load amplitude, and **b** seismic load frequency; DSC_{max} and DSC_{min} represent the maximum and minimum D_{SC}, respectively; and $g = 10$ m/s^2

action time under different seismic load amplitudes or seismic load frequencies. Meanwhile, the reduction of the DSC_{min} of slope caused by the change of the seismic load amplitude is obvious relative to the seismic load frequency, and the DSC_{min} of slope decreases with the increase of the seismic load amplitude or seismic load frequency as a whole. Additionally, the D_{SC} of slope shows a fluctuating trend with the increase of the seismic action time under different seismic load amplitudes or seismic load frequencies, and the fluctuation degree increases and decreases with the increase of the seismic load amplitude and seismic load frequency. Especially, when the seismic load amplitude is 0.02 *g*, 0.04 *g*, 0.08 *g*, 0.10 *g* and 0.12 *g*, respectively, the fluctuation range is 1.2097 ~ 1.7153, 1.0228 ~ 2.0961, 0.9146 ~ 3.7380, 0.9132 ~ 4.7960 and 0.9109 ~ 6.8602; when the seismic load frequency is 3 Hz, 5 Hz, 7 Hz, 8 Hz and 10 Hz, respectively, the fluctuation range is 0.9199 ~ 5.0078, 0.9132 ~ 4.7960, 0.9122 ~ 2.7487, 0.9111 ~ 2.3953 and 0.9097 ~ 2.1408.

It can be seen from Table 3.5 that according to the $ASCM_{min}$ (as shown in Eq. (3.24)), when the seismic load amplitude is 0.02 *g*, 0.04 *g*, 0.08 *g*, 0.10 *g* and 0.12 *g*, respectively (the S_{SC} of slope is 1.74 at this time), the values of the final dynamic stability evaluation index of slope are 1.3008, 1.1793, 1.1090, 1.1081 and 1.1066, respectively. Meanwhile, when the seismic load frequency is 3 Hz, 5 Hz, 7 Hz, 8 Hz and 10 Hz, respectively (the S_{SC} of slope is 1.74 at this time), the values of the final dynamic stability evaluation index of slope are 1.1541, 1.1081, 1.0678, 1.0154 and 0.9687, respectively. Additionally, the final dynamic stability of slope decreases with the increase of seismic load amplitude or seismic load frequency, and the variation characteristics of the nephogram of the maximum shear strain increment of slope under different dynamic characteristic parameters have obvious differences and reasonable regularity.

3.6 Instantaneous D_{SC} of the Rock Slope with the Penetrating Saw-Tooth Rock Joints

3.6.1 Specific Solving Equation of the Coefficient for the Typical Slope

In conclusion, based on the dynamic test and in combination with the existing relevant theories, a dynamic algorithm for solving the D_{SC} of slope considering the strength vibration degradation effect of the rock joints was proposed in this study. Then, the dynamic operation of the algorithm was realized by taking the dynamic stability analysis of the typical rock slope with the penetrating straight rock joints as an example. Subsequently, the parameter analysis of the influencing factors of the dynamic stability was further carried out after the feasibility was verified by the case study. On this basis, the analytical solution of the instantaneous D_{SC} of the typical rock slope with the penetrating saw-tooth rock joints will be tentatively derived below to enrich relevant theories in this field, especially to lay a theoretical foundation and provide useful reference for the dynamic numerical calculation process in this case.

According to the LET and FPM (Zhang & Chen, 2001; Cai et al., 2002; Cheng et al., 2007; Stianson et al., 2015; Liu et al., 2015; Li et al., 2021; Yang et al., 2021), when the slope slides along two (more than two) planes or other forms of broken lines and irregular curves, resulting in ISF, the vertical line can be used to divide the sliding area (sliding wedge) into several rock strips. For each rock strip, the reaction force generated by adjacent rock strips is considered, and the force polygon of each rock strip is drawn (strip diagram). If the polygon is closed when the last force polygon is drawn, it indicates that the slope is just in the limit equilibrium state; otherwise, the slope is in the stable or unstable state (it needs to be determined according to the specific situation). Based on this, combined with Fig. 3.20, the upper slip mass with the penetrating saw-tooth rock joints as the sliding surface was divided into blocks $i-1$, i and $i+1$ ($i \geq 2$); then, the mechanical analysis was carried out on the first two blocks to establish the limit equilibrium equations of the two in the horizontal direction at any moment of seismic duration, while the mechanical analysis was conducted on the third block to establish the limit equilibrium equation on the sliding surface at any moment of seismic duration. Subsequently, the existing strength reduction theory (Liu, 2010; Tu et al., 2016; Sun et al., 2020) was used to further solve the instantaneous D_{SC} of slope, and the specific derivation process is as follows.

For block $i-1$, the limit equilibrium equation in the horizontal direction at any moment of seismic duration is shown in Eq. (3.25). In Eq. (3.25): at any moment t of seismic duration, $R_x(t)$ is the reaction force of block i against $i-1$, $F_{h_{i=1}}(t)$ is the horizontal SIF of block $i-1$, and $R_{i-1}(t)$ is the reaction force of the rock mass at the lower part of block $i-1$ against its sliding surface; l_{i-1} is the length (area per unit width) of the sliding surface of block $i-1$, and θ_{i-1} is the inclination angle of the sliding surface of block $i-1$; in addition, the physical meanings of

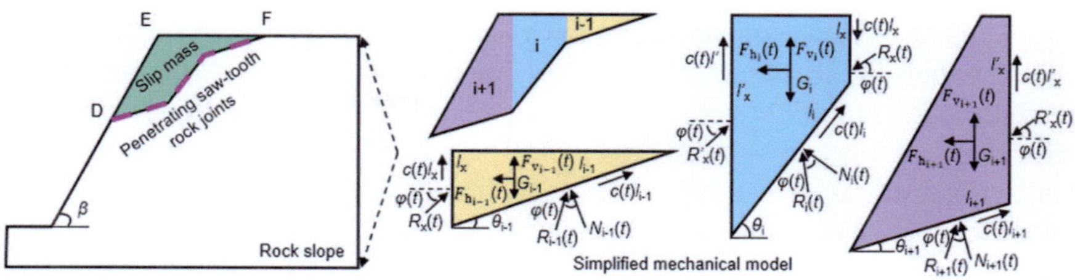

Fig. 3.20 Simplified mechanical model of the typical rock slope with the penetrating saw-tooth rock joints subjected to seismic loading (see Sect. 3.6.1 for the specific meanings of the symbols in this figure)

other symbols are the same as those above-mentioned. Meanwhile, Eq. (3.26) can be obtained through basic mechanical analysis. In Eq. (3.26): at any moment t of seismic duration, $N_{i-1}(t)$ is the normal force outside the sliding surface of block $i-1$, and $F_{v_{i-1}}(t)$ is the vertical SIF of block $i-1$; G_{i-1} is the self-gravity of block $i-1$; in addition, the physical meanings of other symbols are the same as those above-mentioned. Therefore, combined with Eqs. (3.25) and (3.26), the limit equilibrium equation of block $i-1$ in its horizontal direction at any moment of seismic duration can be obtained, as shown in Eq. (3.27), and the physical meanings of all symbols are the same as those above-mentioned in Eq. (3.27).

$$R_x(t)\cos\varphi(t) + c(t)l_{i-1}\cos\theta_{i-1} + R_{i-1}(t)\sin\left[\theta_{i-1} + \varphi(t)\right] = F_{h_{i-1}}(t) \tag{3.25}$$

$$\begin{aligned} R_{i-1}(t) &= N_{i-1}(t)\sec\varphi(t) \\ &= \left[G_{i-1}\cos\theta_{i-1} - F_{h_{i+1}}(t)\sin\theta_{i-1} - F_{v_{i-1}}(t)\cos\theta_{i-1}\right]\sec\varphi(t) \end{aligned} \tag{3.26}$$

$$\begin{aligned} &R_x(t)\cos\varphi(t) + c(t)l_{i-1}\cos\theta_{i-1} + \left[G_{i-1}\cos\theta_{i-1} - F_{h_{i-1}}(t)\sin\theta_{i-1}\right. \\ &\quad \left. - F_{v_{i-1}}(t)\cos\theta_{i-1}\right]\sec\varphi(t)\sin\left[\theta_{i-1} + \varphi(t)\right] = F_{h_{i-1}}(t) \end{aligned} \tag{3.27}$$

For block i, the limit equilibrium equation in the horizontal direction at any moment of seismic duration is shown in Eq. (3.28). In Eq. (3.28): at any moment t of seismic duration, $R'_x(t)$ is the reaction force of block $i+1$ against i, $F_{h_i}(t)$ is the horizontal SIF of block i, and $R_i(t)$ is the reaction force of the rock mass at the lower part of block i against its sliding surface; l_i is the length (area per unit width) of the sliding surface of block i, and θ_i is the inclination angle of the sliding surface of block i; in addition, the physical meanings of other symbols are the same as those above-mentioned. Meanwhile, Eq. (3.29) can be obtained through basic mechanical analysis. In Eq. (3.29): at any moment t of seismic duration, $N_i(t)$ is the normal force outside the sliding surface of block i, and $F_{v_i}(t)$ is the vertical SIF of block i; G_i is the self-gravity of block i; in addition, the physical meanings of other symbols are the same as those above-mentioned. Therefore, combined with Eqs. (3.28) and (3.29), the limit equilibrium equation of block i in its horizontal direction at any moment of seismic duration can be obtained, as shown in Eq. (3.30), and the physical meanings of all symbols are the same as those above-mentioned in Eq. (3.30). Furthermore, combined with Eqs. (3.27) and (3.30), the equation of total force between rock strips transmitted by blocks $i-1$ and i to block $i+1$ at any moment of seismic duration can be obtained, as shown in Eq. (3.31), and the physical meanings of all symbols are the same as those above-mentioned in Eq. (3.31).

$$R'_x(t)\cos\varphi(t) + c(t)l_i\cos\theta_i + R_i(t)\sin[\theta_i + \varphi(t)] = F_{h_i}(t) + R_x(t)\cos\varphi(t) \tag{3.28}$$

$$R_i(t) = N_i(t)\sec\varphi(t) = \left[G_i\cos\theta_i - F_{h_i}(t)\sin\theta_i - F_{v_i}(t)\cos\theta_i\right]\sec\varphi(t) \tag{3.29}$$

$$\begin{aligned} & R'_x(t)\cos\varphi(t) + c(t)l_i\cos\theta_i + \left[G_i\cos\theta_i - F_{h_i}(t)\sin\theta \right. \\ & \quad \left. - F_{v_i}(t)\cos\theta_i\right]\sec\varphi(t)\sin[\theta_i + \varphi(t)] = F_{h_i}(t) + R_x(t)\cos\varphi(t) \end{aligned} \tag{3.30}$$

$$R'_x(t) = \left\{ \begin{array}{l} F_{h_{i-1}}(t) - c(t)l_{i-1}\cos\theta_{i-1} - \left[G_{i-1}\cos\theta_{i-1} - F_{h_{i-1}}(t)\sin\theta_{i-1} \right. \\ \left. - F_{v_{i-1}}(t)\cos\theta_{i-1}\right]\sec\varphi(t)\sin\left[\theta_{i-1} + \varphi(t)\right] \\ + F_{h_i}(t) - c(t)l_i\cos\theta_i - \left[G_i\cos\theta_i - F_{h_i}(t)\sin\theta_i \right. \\ \left. - F_{v_i}(t)\cos\theta_i\right]\sec\varphi(t)\sin[\theta_i + \varphi(t)] \end{array} \right\} \Big/ \cos\varphi(t) \tag{3.31}$$

For block $i+1$, the limit equilibrium equation on the sliding surface at any moment of seismic duration is shown in Eq. (3.32). In Eq. (3.32): at any moment t of seismic duration, $F_{v_{i+1}}(t)$ is the vertical SIF of block $i+1$, $F_{h_{i+1}}(t)$ is the horizontal SIF of block $i+1$, and $R_{i+1}(t)$ is the reaction force of the rock mass at the lower part of block $i+1$ against its sliding surface; l_{i+1} is the length (area per unit width) of the sliding surface of block $i+1$, θ_{i+1} is the inclination angle of the sliding surface of block $i+1$, and l'_x is the adjacent interface length (area per unit width) of block $i+1$ and i; in addition, the physical meanings of other symbols are the same as those above-mentioned. Meanwhile, Eq. (3.33) can be obtained through basic mechanical analysis. In Eq. (3.33): $N_{i+1}(t)$ is the normal force outside the sliding surface of block $i+1$ at any moment t of seismic duration, and G_{i+1} is the self-gravity of block $i+1$; in addition, the physical meanings of other symbols are the same as those above-mentioned. Therefore, combined with Eqs. (3.32) and (3.33), the limit equilibrium equation of block $i+1$ in its sliding surface at any moment of seismic duration can be obtained, as shown in Eq. (3.34). In Eq. (3.34): $R'_x(t)$ is determined by Eq. (3.31); in addition, the physical meanings of other symbols are the same as those above-mentioned.

$$\begin{aligned} & c(t)l_{i+1} + F_{v_{i+1}}(t)\sin\theta_{i+1} + R_{i+1}(t)\sin\varphi(t) \\ & + \left\{R'_x(t)\sin\left[\varphi(t) - \theta_{i+1}\right] - c(t)l'_x\cos\theta_{i+1}\right\}\tan\varphi(t) \\ & \quad = F_{h_{i+1}}(t)\cos\theta_{i+1} + G_{i+1}\sin\theta_{i+1} + c(t)l'_x\sin\theta_{i+1} + R'_x(t)\cos\left[\varphi(t) - \theta_{i+1}\right] \end{aligned} \tag{3.32}$$

$$\begin{aligned} R_{i+1}(t) &= N_{i+1}(t)\sec\varphi(t) \\ &= \left[G_{i+1}\cos\theta_{i+1} - F_{h_{i+1}}(t)\sin\theta_{i+1} - F_{v_{i+1}}(t)\cos\theta_{i+1}\right]\sec\varphi(t) \end{aligned} \tag{3.33}$$

$$\begin{aligned} & c(t)l_{i+1} + F_{v_{i+1}}(t)\sin\theta_{i+1} + \left\{G_{i+1}\cos\theta_{i+1} - F_{h_{i+1}}(t)\sin\theta_{i+1}\right. \\ & \left. -F_{v_{i+1}}(t)\cos\theta_{i+1} + R'_x(t)\sin\left[\varphi(t) - \theta_{i+1}\right] - c(t)l'_x\cos\theta_{i+1}\right\}\tan\varphi(t) \\ & = F_{h_{i+1}}(t)\cos\theta_{i+1} + G_{i+1}\sin\theta_{i+1} + c(t)l'_x\sin\theta_{i+1} + R'_x(t)\cos\left[\varphi(t) - \theta_{i+1}\right] \end{aligned} \tag{3.34}$$

In conclusion, to further solve the instantaneous D_{SC} of slope with the penetrating saw-tooth rock joints, multiple trial calculations are required. Generally, an instantaneous D_{SC} can be assumed, and then the ratio of the shear strength characteristic parameters (cohesion and internal friction coefficient at any moment of seismic duration) of the rock joints to the assumed coefficient can be calculated. Subsequently, Eq. (3.37) can be obtained after the new shear strength characteristic parameters (new cohesion and internal friction angle are shown in Eqs. (3.35) and (3.36), respectively) of the rock jointse gained by the above calculation are brought into Eq. (3.34). If the left and right sides of Eq. (3.37) are equal, the assumed coefficient is the instantaneous D_{SC} of slope; otherwise, the assumed coefficient is continuously updated and the above calculation process is repeated until the required results are obtained (Cai et al., 2002). In Eqs. (3.35)–(3.37): $F_{si}(t)$ is the assumed instantaneous D_{SC} that constantly updated by the numerical program, $\varphi(t)$ and $c(t)$ are determined by Eqs. (3.13) and (3.14), respectively, $R'_x(t)$ is determined by Eq. (3.31); in addition, the physical meanings of other symbols are the same as those above-mentioned.

$$\tan\varphi(t)_{new} = \tan\varphi(t)\big/F_{si}(t) \Rightarrow \varphi(t)_{new} = \arctan\left[\tan\varphi(t)\big/F_{si}(t)\right] \tag{3.35}$$

$$c(t)_{new} = c(t)\big/F_{si}(t) \tag{3.36}$$

$$\begin{aligned} & c(t)_{new}l_{i+1} + F_{v_{i+1}}(t)\sin\theta_{i+1} \\ & +\left\{G_{i+1}\cos\theta_{i+1} - F_{h_{i+1}}(t)\sin\theta_{i+1} - F_{v_{i+1}}(t)\cos\theta_{i+1}\right. \\ & \quad \left. +R'_x(t)\sin\left[\varphi(t)_{new} - \theta_{i+1}\right] - c(t)_{new}l'_x\cos\theta_{i+1}\right\}\tan\varphi(t)_{new} \\ & \quad = F_{h_{i+1}}(t)\cos\theta_{i+1} + G_{i+1}\sin\theta_{i+1} + c(t)_{new}l'_x\sin\theta_{i+1} \\ & \qquad +R'_x(t)\cos\left[\varphi(t)_{new} - \theta_{i+1}\right] \end{aligned} \tag{3.37}$$

3.6.2 Current Research Achievements and Future Prospects

Generally, the dynamic response process of the rock slope subjected to seismic loading reflects the complex mechanical state of the coupling action of the internal and external influencing factors to a certain extent. Among them, the internal influencing factors often refer to the geological structure model of slope and the strength characteristic parameters of the rock joints, etc.; while the external influencing factors usually refer to the excitation of seismic load, etc. The effects of the internal and external influencing factors on the dynamic stability of slope are closely related and interacted. Most of the existing studies have ignored the close relationship between the internal and external influencing factors, and rarely considered the impact of the excitation of seismic load on the shear strength properties of the rock joints of slope. In view of this, this study established the equation for solving the SVDC of the rock joints, and regarded it as a bridge between the internal and external influencing factors; then, quantitatively analyzed the impact of the excitation of seismic load on the strength characteristic parameters of the rock joints of slope, and thus deeply explored the dynamic stability of slope from the perspective of the coupling action of the internal and external influencing factors. In particular, on the basis of existing studies, this study summarized the main factors that affect the strength vibration degradation effect of the rock joints as the vibration wear and relative motion velocity between rock blocks. Among them, the equation for solving the influence factor of vibration wear was determined by conducting the cyclic shear test of the penetrating rock joints, and the equation for solving the influence factor of relative motion velocity refered to the previous research results.

It needs to be clarified that although the lithology, loading state and shape of the rock joints used in the cyclic shear test are different slightly from those in the existing studies at present, the change rules revealed by that can still be used for reference in this study. Therefore, the research conclusions obtained in this study also have certain theoretical significance. Especially, the relevant research carried out in this study is still at the preliminary exploration stage. On the one hand, the classification of the influencing factors of the strength vibration degradation effect of the rock joints, and the correlation between the internal and external influencing factors need to be further clarified via more dynamic tests. On the other hand, in the stability analysis of slope under the excitation of actual seismic load, the application of the strength vibration degradation rules of the rock joints revealed by the cyclic shear test needs to be further verified by more examples. In the future, efforts will be made to obtain a more universal equation for solving the SVDC of the rock joints, so as to provide a reliable theoretical basis for the long-term dynamic stability analysis and evaluation of the rock slope.

3.7 Summary and Conclusions

(1) The shear strength vibration degradation effect of the rock joints is significantly affected by the vibration wear (irreversible) and relative motion velocity (reversible) between rock blocks. The dynamic change of the SVDC of the rock joints with seismic time can be introduced to quantitatively describe the degradation effect of the above factors on the strength of the rock joints during

the seismic duration, and this coefficient is a dynamic variable that depends on the response value of the number of cycles, cyclic shear amplitude and relative motion velocity between rock blocks.

(2) The analytical solution of the SVDC of the rock joints affected by multiple factors was firstly obtained via the dynamic test; then, this analytical solution was embedded into the equation of the D_{SC} of slope combined with the LET and DVM; subsequently, the built-in FISH language of the UDEC software was used to program and refresh the strength characteristic parameters of the rock joints in real-time, and capture the SIF of slope at any moment of seismic duration; moreover, the $ASCM_{min}$ was used to solve the value of the final dynamic stability evaluation index of slope; thereby, a dynamic algorithm of the D_{SC} of slope considering the strength vibration degradation effect of the rock joints was proposed.

(3) Taking the dynamic stability analysis of the rock slope with the penetrating straight rock joints as an example, the simple harmonic wave was used to simulate the seismic load, and the response characteristics of the dynamic stability of slope were analyzed by using the algorithm proposed in this study. The results show that the calculation results obtained by considering the strength vibration degradation effect of the rock joints when solving the D_{SC} of slope are more in line with the general natural rules. Meanwhile, the algorithm proposed was applied to the case study of the dynamic stability of slope under the background of Wenchuan seismic wave, the analysis results are consistent with the shaking table model test results, which further verifies the feasibility of the algorithm. Therefore, in the analysis and evaluation of the dynamic stability of slope, the influence of the strength vibration degradation effect of the rock joints on its stability should not be ignored.

(4) Under the influence of different strength characteristic parameters the rock joints, geometric parameters of slope and dynamic characteristic parameters, the $SVDC_{min}$ of the rock joints decays constantly in the form of negative exponential function with the variation of dynamic excitation time. When the initial cohesion (initial internal friction angle) of the rock joints is larger, and the slope angle (dip angle of bedding plane) or seismic load amplitude (seismic load frequency) is smaller, the DSC_{min} of slope is larger during the seismic duration, and the final dynamic stability of slope is relatively stronger at this time. Furthermore, the method of solving the instantaneous D_{SC} of the rock slope with the penetrating saw-tooth rock joints was discussed, and the current research achievements and future prospects were expounded, which is expected to provide a reliable theoretical basis for the long-term dynamic stability analysis and evaluation of the rock slope.

Chapter 4
Dynamic Response Rules and Damage Mechanism of the BRS Under Deterioration of Rock Mass in the HB

4.1 Introduction

The TGP alleviates the unfavorable situation of power supply shortages in most areas of China, ensures the safe control of the water-level in the areas along the Yangtze River during the flooding season, and generally promotes the rapid development of the shipping economy in the Yangtze River (Liu et al., 2013; Li et al., 2021; Deng et al., 2022). However, after the project was completed and put into operation, the original geological environment in the area affected by the long and narrow reservoir (covering an area of thousands of square kilometres) has been changed or destroyed, resulting in frequent REs (mainly microseismic waves and low seismic waves) and gradual deterioration of rock mass in the HB (Gu & Huang, 2016; Shang et al., 2019; Tang et al., 2019). Hence, tens of thousands of secondary geological disasters (bedding landslides are relatively common) have been triggered, such as landslides, collapses, mud-rock flows, swells, ground collapses and ground fissures (Fu et al., 2010; Huang et al., 2010; Deng et al., 2017; Wang et al., 2019; Du et al., 2022; Xiong et al., 2022). Tens of millions of residents in the TGRA have been affected by these disasters (Fourniadis et al., 2007; Yin et al., 2016; Xu et al., 2022a, c), and more attention should be given to these phenomena. Therefore, to offer a useful reference for comprehensively improving the disaster prevention, mitigation and relief capabilities in the TGRA, it is extremely necessary to conduct a study on the dynamic response rules and damage mechanism of the BRSs subjected to frequent REs when taking into account the deterioration of rock mass in the HB.

At present, most scholars have unilaterally conducted studies on the impact of either the REs or the deterioration of rock mass in the HB on the stability of the bank slopes in the TGRA. On the one hand, the number of the REs dominated by microseismic waves and low seismic waves has continued to increase since impoundment in the TGRA, especially in the region of Badong County → Zigui County → Wushan County (Yao et al., 2017; Li et al., 2019; Tang et al., 2019; Xu et al., 2022a, c), which has led to frequent landslides on the BRSs in this area. In particular, the existing

X. Liu and B. Xu, *Dynamic Stability of the Bedding Rock Slope with Deteriorated Hydro-fluctuation Belt*, https://doi.org/10.1007/978-981-95-6311-1_4

studies (Yin et al., 2010; Huang, 2018; Xu et al., 2022a, c) noted that although the seismic wave energy released by a few REs does not directly damage the bank slope and cause it to lose stability, a series of sudden secondary geological disasters often occur on the bank slope after the cumulative damage caused by perennial REs, and the processes of gestation → development → outbreak → stability of such disasters are often highly coherent and irresistible. Focusing on this theme, the earlier study of Chen and Talwani (1998) showed that the REs generally has the characteristics of small magnitude, shallow seismic source, high frequency of occurrence, strong epicentral intensity and rapid intensity attenuation. Subsequently, Ma et al. (2010) accurately located nearly 3000 microseismic waves and low seismic waves with magnitudes of 0.8 ~ 2.9 in the TGRA, and analyzed the causes of occurrence. Guo et al. (2012) analyzed the characteristics of the spatial distribution of seismic sources before and after impoundment, and the reasons for the change in seismic frequency based on the data for the REs in the TGRA since 1980. Jiang et al. (2016), Liu et al. (2018), Deng et al. (2020a, b), Yang et al. (2020) and Xu et al. (2022a, c) employed the shaking table model tests and (or) UDEC numerical explorations to systematically assess the evolution mechanism of the DCD and ISF, and the factors influencing the stability of different types of bank slopes subjected to multiple microseismic waves and low seismic waves. The results of the above studies enriched the relevant theories of the dynamic stability analysis and predictions of the bank slopes in an environment of frequent microseismic waves and low seismic waves in the TGRA, and improved upon the limitations of most studies that only considered the action of high seismic waves or a few microseismic waves (low seismic waves). On the other hand, the rock mass in the HB of the TGRA continues to deteriorate and undergo damage under the impact of perennial periodic fluctuations of the reservoir water-level, and increasing irreversible degradation has already caused a hidden danger to the stability of the upper bank slope that can not be underrated (Wang et al., 2017; Huang et al., 2021; Xu et al., 2022a, c). Focusing on this theme, Wang (2006) previously used laboratory tests, theoretical analysis and numerical explorations to investigate the deterioration characteristics of rock mass in the HB of the TGRA under the water–rock interaction, and its impact mechanism on the bank slope's stability. Subsequently, Rinaldi et al. (2020) proposed a method to study the characteristics of the REs related to changes in water-level of the HB. Liang et al. (2021) studied the strength degradation trends of rock mass in the HB of the TGRA by the dry–wet cycle tests and established corresponding fracture criteria. Fang et al. (2021) used seepage tests, joint surface scanning tests and ion concentration measurements to explore the seepage characteristics and degradation mechanism of rock mass in the HB of the TGRA. The achievements of the above studies revealed the evolution characteristics of the deterioration of rock mass in the HB of the TGRA and its regulatory mechanism on the bank slope's stability, which laid a theoretical basis and provides a useful reference for further exploration of the bank slope's stability.

It should be emphasized that only a few scholars have discussed the bank slope's stability in the TGRA under the coupled action of the REs and the deterioration of rock mass in the HB at present. For example, Wang et al. (2018) analyzed and evaluated the seismic performance of the bank slope under full consideration of the deterioration effect of the shear strength parameters of the rock-soil mass of the HB in the TGRA. Zhang et al. (2017, 2018) analyzed the relationship between the water–rock interaction of the HB and frequent REs in the TGRA, as well as the dynamic response features of the bank slope under the coupled action of the two factors. Xu et al. (2022a, c) used a series of the shaking table model tests and UDEC numerical explorations in a preliminary exploration of the evolution processes of the bank slope's damage instability, and the factors that influence its stability while considering the impact of deterioration of rock mass in the HB of the TGRA. In summary, there are few research reports that consider the impact of the coupled action of the REs and the deterioration of rock mass in the HB on the stability of the bank slope, especially the evolution mechanism of the DCD and ISF of the bank slope under this load environment. Thus, it is urgent to conduct an in-depth and systematic study on this key scientific issue.

In view of this, this study described a systematic and deep study of the dynamic response rules and damage mechanism of the BRSs (including gently (GI_S), moderately (MI_S) and steeply (SI_S) inclined slopes) affected by the deterioration of rock mass in the HB based on the previous studies. According to the findings of the shaking table model tests, the response rules of the acceleration, C_D, rock pressure (R_P), pore-water pressure (P_W), damping ratio (D_R) and natural frequency (N_F) of various slopes above-mentioned were analyzed, the unified mathematical models of the nonlinear DCD of slopes were established, and the specified evolution processes of the DCD and ISF of slopes were revealed and generalized. The conclusions of this research can offer a theoretical basis and valuable reference for the prevention and control of geological disasters and the bank slope's long-term stability in the TGRA.

4.2 Generalization of the Slope Model

According to the in-situ investigation and analysis of hydrology and engineering geology in the TGRA described in Sects. 1.2–1.4, fully considering the key influence factors such as the perennial scouring and infiltration of reservoir, the characteristics of bedding planes of slope, the macro-scale morphological features of the HB, the dynamic disaster formation mechanism of the bank slope and the laboratory test conditions, the limestone BRS containing the HB was generalized by a typical slope model as shown in Fig. 4.1 (taking the GI_s as an example, other types of slopes were the same). As a result, the dynamic response mechanism and stability of the typical BRSs subjected to multiple REs, when considering deterioration of rock mass in the HB, were investigated in this study.

Notably, given the similarity design conditions of the shaking table model tests as described in Sect. 4.3.1, the slope height of the slope model generalized in the tests was 1000 mm, the slope width was 400 mm, the slope angle was 60°, the spacing between bedding planes was 50 mm, the spacing between secondary joints was 60 mm, the depth of secondary joints cuts corresponded to 2/3 of the spacing between bedding planes, and the dip angles of bedding planes included 30° (GI_s),

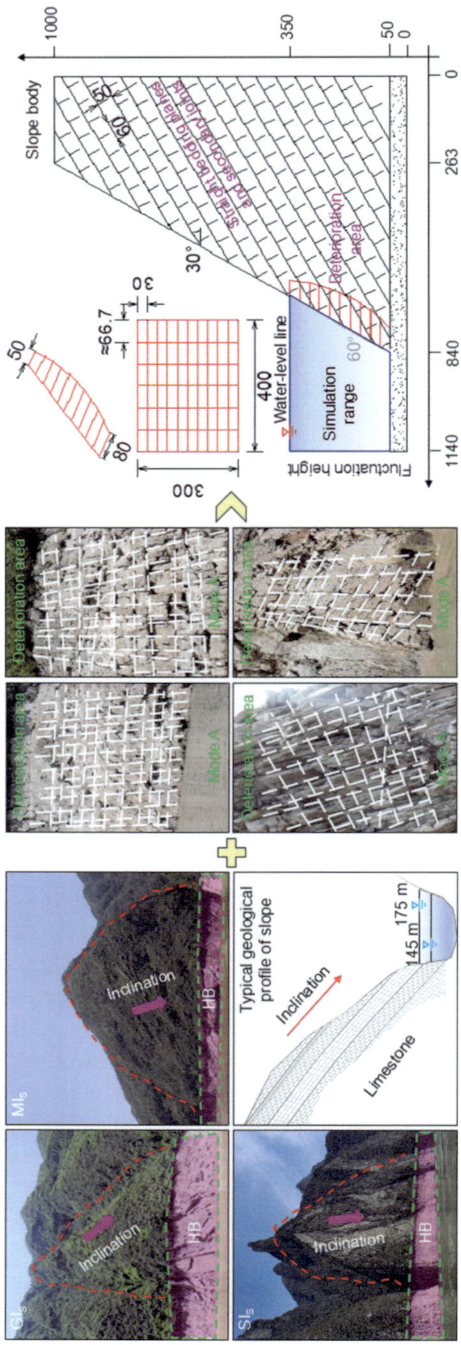

Fig. 4.1 Typical BRSs in the TGRA and the generalization diagram of the slope model (unit: mm)

45° (MI_s) and 60° (SI_s). Additionally, the macro-scale deterioration morphology of rock mass in the HB was characterized by the erosion (denudation) type (as shown in Mode A in Fig. 1.3), and its elevation difference and cutting depth range were 300 mm (reflected by the water-level elevation of 145 ~ 175 m in the reservoir bank) and 50 ~ 80 mm, respectively.

4.3 Shaking Table Model Test

4.3.1 Similarity Design

Based on the similarity design theory (Tao et al., 2019; Tao et al., 2021; Zhu et al., 2021; Zang et al., 2022; Xu et al., 2022a, c), following the basic principle of making the model as easy to operate as possible on the premise of ensuring the measurement accuracy of test parameters, and considering the key influence conditions such as main research problems, equipment performance and test cycle, the similarity coefficients of density (ρ), elastic modulus (E) and geometric size (l) were selected as the basic control quantities (BCQs) in the tests (taking $C_\rho = 1$, $C_E = 100$ and $C_l = 100$, respectively). Furthermore, according to the mass—length—time (MLT) dimension system and the π theorem, the similarity coefficients of other important physical attributes were determined by dimensional analysis, as shown in Table 4.1.

Additionally, the arenaceous quartz (coarse aggregate), barite powder (fine aggregate), gypsum powder (cementitious agent), retarder (to extend the initial setting time

Table 4.1 Important physical attributes and their similarity coefficients

Important physical attribute	Dimension	Similarity relation	Similarity coefficient
Density	ML^{-3}	C_ρ	1 (BCQs)
Elastic modulus	$ML^{-1}T^{-2}$	C_E	100 (BCQs)
Geometric size	L	$C_l = {C_E}^{0.5}{C_\rho}^{-0.5}C_t$	100 (BCQs)
Cohesion	$ML^{-1}T^{-2}$	$C_c = C_E C_\varepsilon$	100
Internal friction angle	Dimensionless	C_φ	1
Stress	$ML^{-1}T^{-2}$	$C_\sigma = C_E C_\varepsilon$	100
Strain	Dimensionless	$C_\varepsilon = C_\rho C_g C_l {C_E}^{-1}$	1
Poisson's ratio	Dimensionless	C_μ	1
Displacement	L	$C_u = C_l C_\varepsilon$	100
Time	T	C_t	10
Frequency	T^{-1}	$C_f = {C_t}^{-1}$	0.1
Velocity	LT^{-1}	$C_v = C_u {C_t}^{-1}$	10
Acceleration	LT^{-2}	$C_g = C_u {C_t}^{-2}$	1
Damping ratio	Dimensionless	C_ζ	1

of gypsum powder), glycerol (water retention additive) and tap water were chosen as the starting compounds of the equivalent materials of limestone used in the tests (as shown in Fig. 4.2a) with reference to the equivalent materials frequently employed in the geomechanical model tests (Liu et al., 2018c; Deng et al., 2020a, b; Yang et al., 2020; Xu et al., 2022a, c). There were dozens of attempts with various proportions of the equivalent materials, and the preparation of test specimens of the equivalent materials mainly included batch weighing, stirring evenly, filling and tamping, and demoulding and molding, as shown in Fig. 4.2b). The oven-drying method, unconfined compression tests and direct shear tests (as shown in Fig. 4.2c) were used to determine an optimal proportion (mass ratio) of the equivalent materials, i.e., arenaceous quartz: barite powder: gypsum powder: glycerol: tap water = 1: 0.387: 0.211: 0.037: 0.187, and the retarder content ranged from 0.20% to 0.25% of the gypsum powder's weight in the reaction mixture.

The physical–mechanical parameters (including ρ, compressive strength (σ_c), E, Poisson's ratio (μ), cohesion (c) and internal friction angle (φ)) and relevant test curves of the equivalent materials for the optimal proportion are shown in Table 4.2 and Fig. 4.2d, e, respectively. The prototype values in Table 4.2 were obtained in the previous stage through conventional physical–mechanical property tests on the initial limestone specimens taken from the Wushan segment (as shown in Fig. 1.3) in the TGRA (Liu et al., 2021; Xu et al., 2021; Xu et al., 2022b). Furthermore, the fuzzy comprehensive evaluation method was used to assess the similarity between the actual and prototype values of the equivalent materials, and the calculated comprehensive evaluation index value was 0.8894 (approaching 1), which verified the reliability of the proportion of the equivalent materials of limestone (Xu et al., 2022a, c).

4.3.2 Sensor Arrangement

To evaluate the dynamic response of the acceleration, C_D, R_P and P_W of slope subjected to multiple seismic loads in real-time during the tests, eight acceleration sensors (Nos. A1 ~ A8) and eight R_P sensors (Nos. R1 ~ R8) were arranged at the top, shoulder, surface and inner parts of slope, four C_D sensors (Nos. D1 ~ D4) were placed on the top, shoulder and surface of slope, and three P_W sensors (Nos. P1 ~ P3) were arranged in the HB. It should be emphasized that all types of sensors have a certain distance from the slope top or slope surface, i.e., they were not fully exposed on the slope top or slope surface. The specified arrangement of various sensors (along the vertical middle section of the slope model) is displayed in Fig. 4.3 (taking the GIs as an example, other types of slopes were the same).

Among them, the horizontal distance (HD) between A1 (R1) and A6 (R6) and the HD between R1 and R6 were 131.5 mm, the HD between A4 and A2/A7 and the HD between R4 and R2/R7 were 159.5 mm, the HD between A5 and A3/A8 and the HD between R5 and R3/R8 were 253 mm, and the vertical distance (VD) between A2 (A4, A7) and A1 (A6)/A3 (A5, A8) and the VD between R2 (R4, R7) and R1 (R6)/ R3 (R5, R8) were 325 mm. The HD between D1 and D2 was 131.5 mm, the VD

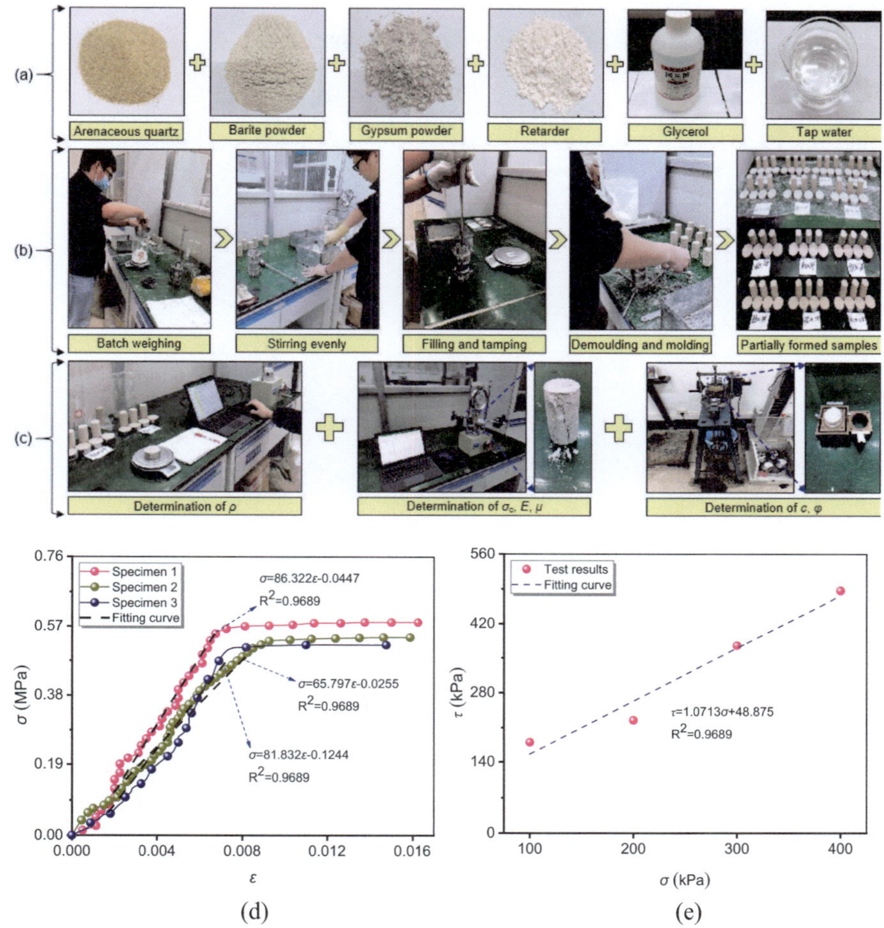

Fig. 4.2 **a** Starting compounds of the equivalent materials, **b** preparation of test specimens of the equivalent materials, **c** tests for assessing the physical–mechanical parameters of the equivalent materials, **d** curve of normal stress (σ) versus axial strain (ε) of the equivalent materials, and **e** curve of shear stress (τ) versus σ of the equivalent materials

Table 4.2 Physical–mechanical parameters of the equivalent materials

Item	Density (g/cm^3)	Compressive strength (MPa)	Elastic modulus (MPa)	Poisson's ratio	Cohesion (kPa)	Internal friction angle (°)
Prototype value	2.65	57.26	6600	0.24	5260	44.53
Similarity coefficient	1	100	100	1	100	1
Target value	2.65	0.57	66.00	0.24	52.60	44.53
Actual value	2.33	0.55	77.98	0.22	48.88	46.97

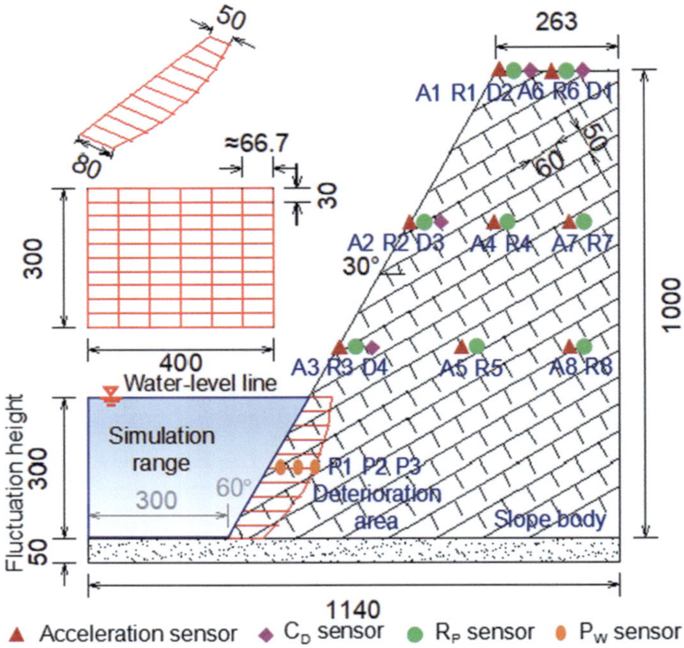

Fig. 4.3 Schematic illustration of sensor arrangement (unit: mm)

between D3 and D2/D4 was 325 mm, and the HD between P2 and P1/P3 was 30 mm. In particular, it was necessary to check whether the sensors were performing correctly to meet the data monitoring standards before embedding or installing various sensors, and fully prepare the tools and materials required for embedding or installing various sensors. In addition, the laying method of "S–type" should be adopted for the lead wires of various sensors, and the length of the lead wires should be guaranteed to have sufficient redundancy to avoid damage and failure of the sensors due to excessive pressure or tension caused by the rock slide and collapse of slope during or after construction of the slope model.

4.3.3 Test Apparatus

The test apparatuses used in this study mainly included a loading system (as shown in Fig. 4.4a), a data acquisition and monitoring system (as shown in Fig. 4.4b), and a model box and water circulation system (as shown in Fig. 4.4c).

(1) The loading system was small indoor shaking table equipment, which was mainly composed of a table top (1.2 m in length and width), hydraulic actuator, oil source, accumulator, servo controller and computer program. The load bearing range of the equipment was 1 t, the operating frequency range was 0

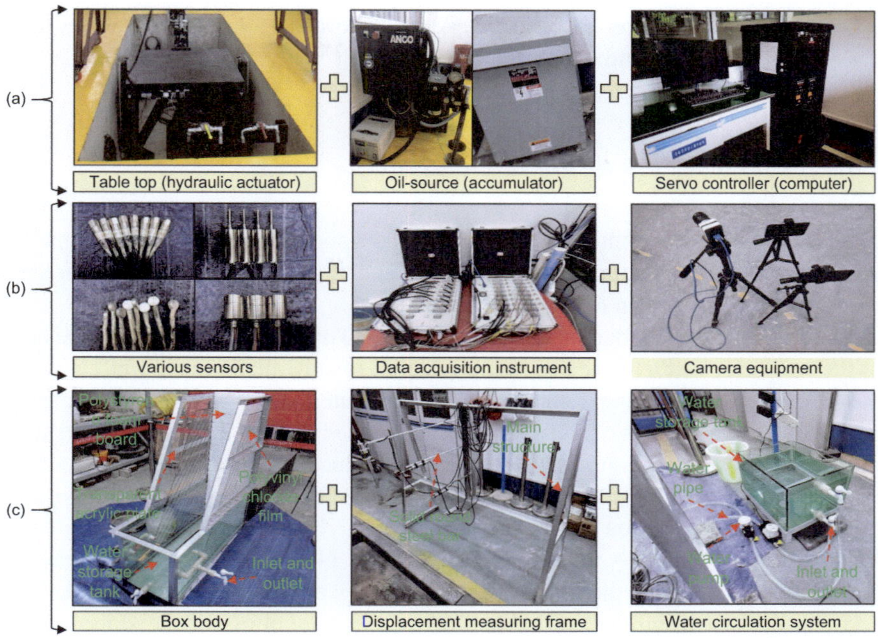

Fig. 4.4 **a** Loading system, **b** data acquisition and monitoring system, and **c** model box and water circulation system

~ 50 Hz, the horizontal displacement range was 100 mm, and the horizontal (vertical) acceleration range was 1.2 *g*/1 t (2 *g*/0.5 t).

(2) The data acquisition and monitoring system mainly included the acceleration sensor, C_D sensor, R_P sensor, P_W sensor, multi-channel static and dynamic data acquisition device, and high-definition camera equipment. The measuring range and operating frequency of the acceleration sensor were ± 10 *g* and 0.6 ~ 5000 Hz, respectively; the measuring range and dynamic response frequency of the C_D sensor were 50 mm and less than 50 Hz, respectively; the measuring range and sensitivity of the R_P sensor were 0.5 MPa and 0.00337 ~ 0.00499 mV/kPa, respectively; and the measuring range and resolution of the P_W sensor were 20 kPa and 0.01 kPa, respectively.

(3) The model box and water circulation system was independently designed and commissioned by the factory. The box body of the model box system primarily consisted of the bottom water storage tank (including steel plate, equilateral angle steel, transparent high-strength tempered glass, and inlet and outlet water valves), side trapezoidal frame (including equilateral angle steel and transparent acrylic plate) and back steel plate connected by ordinary bolts. The displacement measuring frame of the model box system primarily consisted of the main structure (equilateral angle steel) and solid round steel bar connected by common bolts. The water circulation system primarily consisted of the water storage tank (including equilateral angle steel, transparent high-strength tempered glass, and

inlet and outlet water valves), water pump (flow of 15 L/min) and water pipe, which were fixed and connected by pipe clips.

4.3.4 Construction of the Slope Model

The construction of the three typical BRS models above-mentioned was completed in the box body from bottom to top in the tests, and the specified construction process was as follows (as shown in Fig. 4.5).

(1) The polystyrene foam board (thickness of 30 mm) was placed inside the back of the box body as the wave absorbing material to slow the boundary effect of the box body. (2) The polyvinyl chloride film was coated with vaseline (including the profile of bedding planes of slope) and pasted on the inside of the right and left sides of the box body as the lubricating materials to reduce the friction resistance of the box body. (3) The equivalent materials for the slope model were prepared according to the optimal proportion determined in Sect. 4.3.1 (weighing raw materials → mixing dry aggregates → mixing wet aggregates). (4) The foundation cushion that can effectively alleviate the impact of the boundary effect at the bottom of the box body on the test results was filled and tamped at the bottom of the slope model, and the cushion surface was leveled. (5) The bedding planes in the upper part of the slope model were filled and tamped layer by layer, and the bedding plane surface was leveled. (6) The secondary joints and the HB (deterioration area of rock mass) of the slope model were made (including evenly scattering dry fine sand and mica after slow cutting, and the thickness of the materials sprayed on the bedding planes and the width of the joint fissures are approximately 1 mm). (7) Various sensors were buried according to the key points of sensor arrangement in Sect. 4.3.2 (the C_D sensors were placed after the slope model was formed). (8) Static curing and molding were completed.

4.3.5 Loading Conditions

The natural wave (NW) applied in the tests was the wave of the REs captured by an earthquake monitoring station in the TGRA, and the artificial wave and white noise (WN) were the sine wave (SW) and random wave, respectively. Given a series of features of the REs as described in Sects. 1.2 and 4.1, it was often difficult to divide the foreshocks, main shocks and aftershocks in detail; thus, the main seismic wave was generally intercepted for analysis.

Following high-frequency noise reduction and baseline normalization (as shown in Fig. 4.6a), the natural main seismic wave was further spliced (Nos. #1 ~ #10 as shown in Fig. 4.6b), and condensed rendition of the entire seismic waveform sequence through the similarity coefficient of time, with each main seismic wave lasting around 6.5 s. The fluctuation of the main frequency ranged from 0.8 ~ 13.6 Hz after Fourier

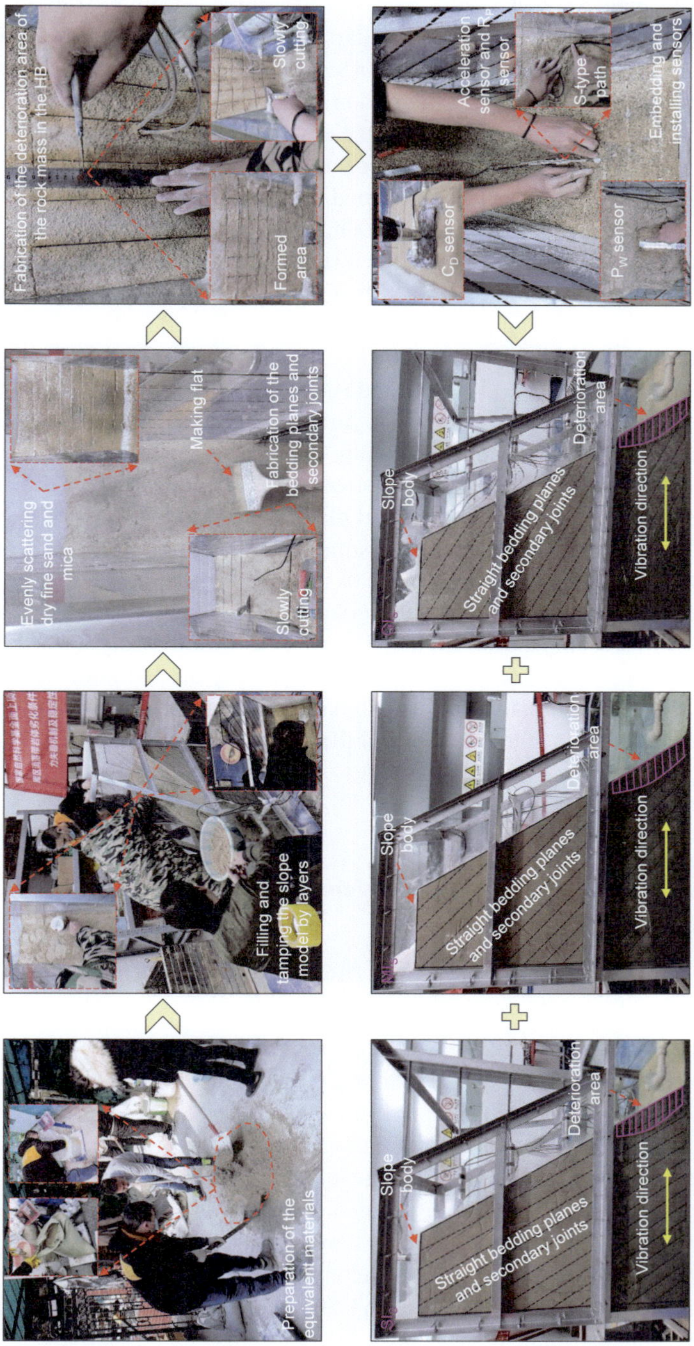

Fig. 4.5 Specified construction processes of the slope model

transformation of the natural main seismic wave (as shown in Fig. 4.6c); thus, the fluctuation of the main frequency employed in the tests ranged from 8 ~ 136 Hz based on the similarity coefficient of frequency. In particular, the available hydrogeological survey data and studies showed that the mean velocity range of the river section was 0.08 ~ 0.45 m/s when the water-level elevation in the TGRA was 175 m (Xu et al., 2022a, c); thus, the mean velocity range of the outlet and inlet in the tests was 0.8 ~ 4.5 cm/s based on the similarity coefficient of velocity.

Table 4.3 shows the specified loading conditions determined in the tests based on the previous studies (Liu et al., 2018c; Deng et al., 2020a, b; Yang et al., 2020; Xu et al., 2022a, c), and especially the operating condition number mentioned in Table 4.3 refers to the cumulative loading times of seismic waves. Among them, the duration and frequency of the main seismic wave were 6.5 s and 30 Hz, respectively, and its amplitude included 0.04, 0.08, 0.10, 0.12, 0.15 and 0.35 *g*. The SW's duration was 6.5 s, its frequency included 5, 15, 20 and 30 Hz, and its amplitude included 0.05, 0.08, 0.10, 0.12, 0.15 and 0.40 *g*. The WN's duration was 30 s, with an amplitude of 0.06 *g*. It should be emphasized that the loading direction of the seismic waves in the tests was horizontal. In particular, the high-strength bolts were employed to transfer via the pre-set bolt holes on the bottom steel plate to closely link the box body and the shaking table as a whole, to guarantee simultaneous motion of the pair during the tests. Meanwhile, the average velocities of the inlet and outlet were maintained at approximately 4.42 cm/s, and the water circulation system continued to operate until the loading was completely completed in the tests. Additionally, the evolution characteristics of the DCD and ISF of slope could be captured, photographed and documented in live action via the clear and durable tempered glass (acrylic plate) panels situated on the right and left flanks of the box body, or on the frontal (upper) facade of the box body.

4.4 Model Test Results and Analysis

4.4.1 Dynamic Response Rules of the Acceleration of Slope

Figure 4.7a, c, e display the variation curves of the amplification coefficient of peak-ground acceleration (PGA) of each monitoring point on the horizontal path (A2 → A4 → A7 and A3 → A5 → A8) and vertical path (A1 → A2 → A3 and A6 → A7 → A8) with the relative width *b*/*B* and relative height *h*/*H* after 500 microseismic waves on the three typical BRSs above-mentioned (i.e., to explore the response rules of different paths). Figure 4.7b, d, f demonstrate the variation curves of the PGA amplification coefficient of each vertical monitoring point of the surface (A1 ~ A3) and inner of slope (A6 ~ A8) with the relative height *h*/*H* after the three typical BRSs above-mentioned underwent initiation ~ 500 microseismic waves (Stage DA1) → 500 ~ 1000 microseismic waves (Stage DA2) → epicenter ~ 500 low seismic waves (Stage DA3) → 500 ~ 1000 low seismic waves (Stage DA4) → 100 high

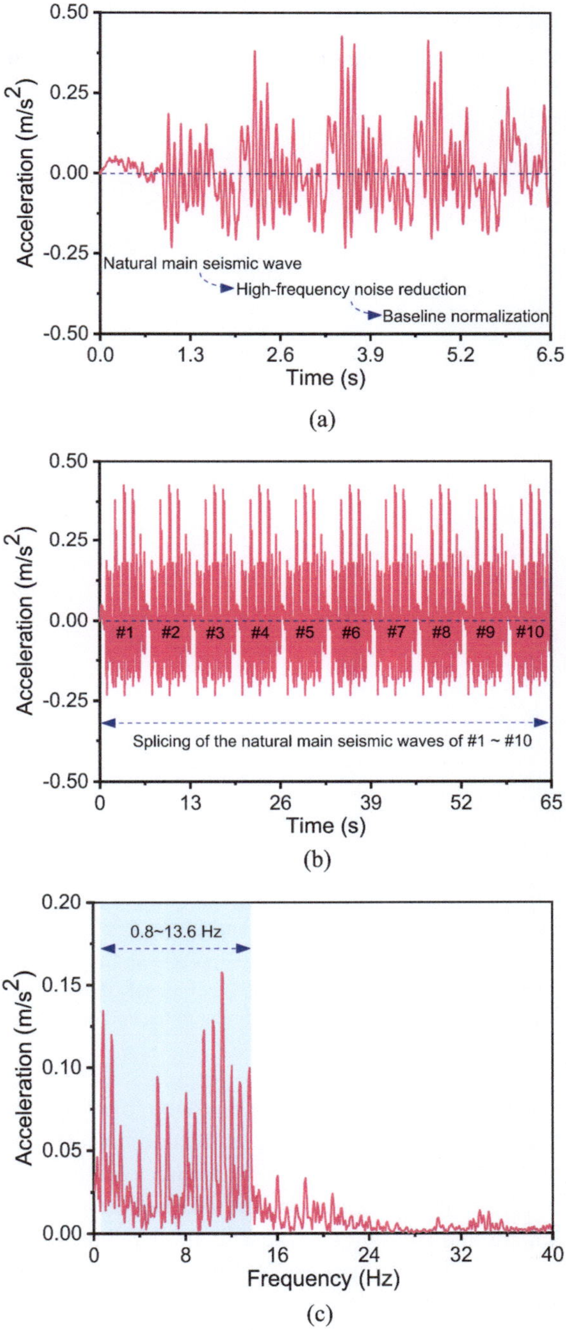

Fig. 4.6 Variation curves of the acceleration temporal profile and fourier amplitude spectrum distribution: **a** high-frequency noise reduction and baseline normalization of the natural main seismic wave, **b** splicing processing of the natural main seismic wave, and **c** fourier amplitude spectrum

Table 4.3 Specified operating conditions of the shaking table model tests

Operating condition number (Cumulative loading times)	Type	Duration (s)	Frequency (Hz)	Amplitude ($\times g$)	Loading times
1#	WN1	30	–	0.06	1
2#	NW	6.5	30	0.04	1
3#	SW	6.5	30	0.05	1
4# ~ 8#	NW	6.5	30	0.04 ~ 0.15	5
9# ~ 12#	SW	6.5	5 ~ 30	0.05	4
13#	WN2	30	–	0.06	1
14# ~ 63#	NW	6.5	30	0.04	50
64#	WN3	30	–	0.06	1
…	Repeat operating conditions 14# ~ 64# until 1000 microseismic waves				
1034# (High intensity in epicenter)	NW	6.5	30	0.35	1
1035#	WN23	30	–	0.06	1
1036# ~ 1085#	SW	6.5	30	0.08 ~ 0.12	50
1086#/1137#/ 1188#/1239#	WN24 ~ 27	30	–	0.06	4
1240# ~ 1289#	SW	6.5	30	0.15	50
1290#	WN28	30	–	0.06	1
…	Repeat operating conditions 1240# ~ 1290# until 1000 low seismic waves				
2056# ~ 2075#	SW	6.5	30	0.40	20
2076#	WN44	30	–	0.06	1
…	Repeat operating conditions 2056# ~ 2076# until 100 high seismic waves				
The SW with a duration of 6.5 s, a frequency of 30 Hz and an amplitude of 0.40 g is continuously loaded until the slope is completely destroyed, and the water circulation system continues to operate from the beginning to the end of loading					

seismic waves (Stage DA5) (i.e., to explore the response rules of different stages). In particular, the PGA amplification coefficient was defined as the ratio of the peak ground acceleration at each monitoring point to the reference point (Xu et al., 2022a, c). Meanwhile, b represents the HD between A2, A4, A7 and A2 or between A3, A5, A8 and A3, and B represents the HD between A7 and A2 or between A8 and A3, where the reference points are A2 and A3; h represents the VD between A1 ~ A3 and A3 or between A6 ~ A8 and A8, and H represents the VD between A1 and A3 or between A6 and A8, where the reference points are A3 and A8.

It can be seen from Figs. 4.7a, c, e that the PGA amplification coefficient of slope gradually increased along the A7 → A4 → A2 and A8 → A5 → A3 paths (from the inner to the surface of slope), and its amplitude changed from steep to gentle with the

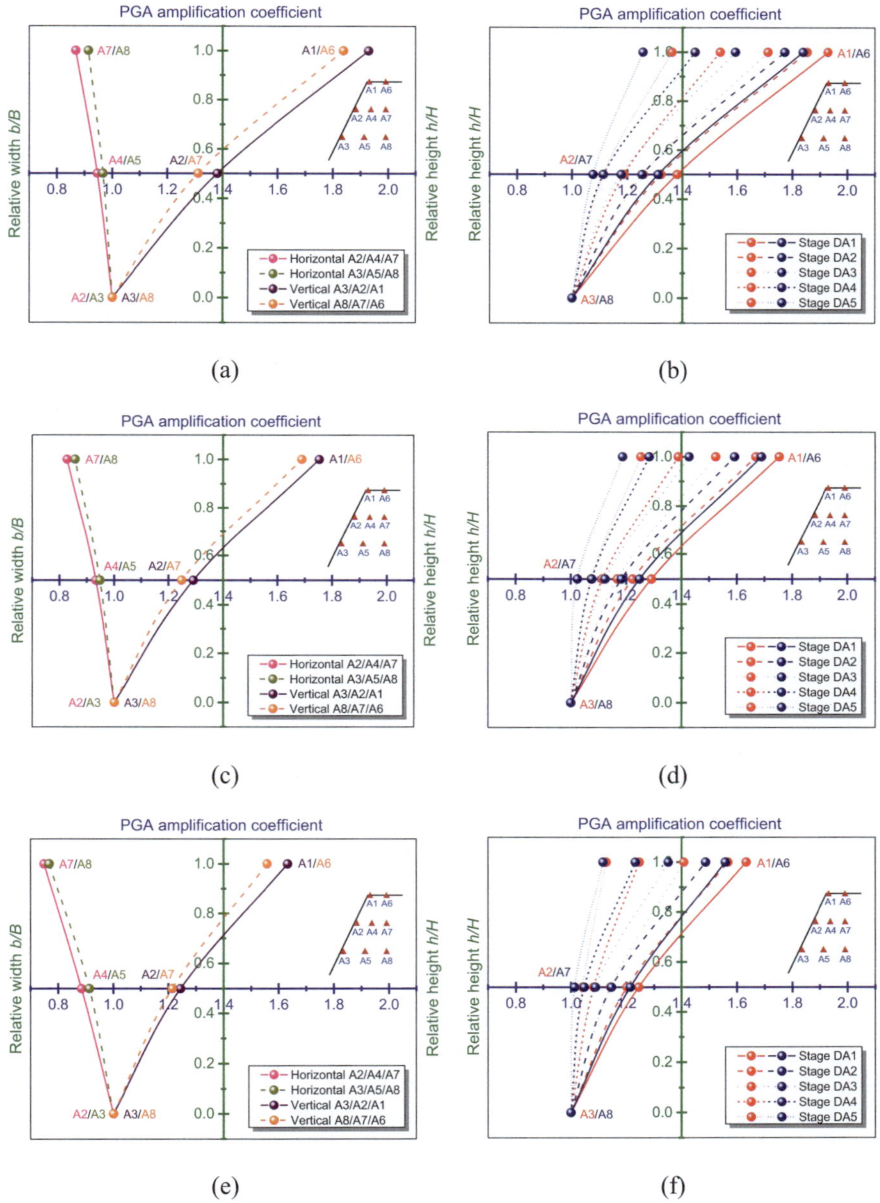

Fig. 4.7 Variation curves of the PGA amplification coefficient of slope: **a** GI_S—different paths, **b** GI_S—different stages, **c** MI_S—different paths, **d** MI_S—different stages, **e** SI_S—different paths, and **f** SI_S—different stages; Stages DA1, DA2, DA3, DA4 and DA5 represent "initiation ~ 500 microseismic waves", "500 ~ 1000 microseismic waves", "epicenter ~ 500 low seismic waves", "500 ~ 1000 low seismic waves" and "100 high seismic waves", respectively

gradual increase in *b*/*B*, which indicated that the acceleration response of slope had an obvious "Surface effect" (the free surface's amplification features) in the horizontal direction. Meanwhile, the PGA amplification coefficient of slope showed a gradually increasing trend along the A3 → A2 → A1 and A8 → A7 → A6 paths (from the base to the crest of slope), and its amplitude changed from steep to gentle with the gradual increase in *h*/*H*, which indicated that the acceleration response of slope had a significant "Elevation effect" (the vertical direction's amplification characteristics) in the vertical direction. In particular, the PGA amplification coefficients of the monitoring points located at the slope shoulder (A1) and the slope top (A6) were remarkably greater than that of the monitoring points located on the slope surface (A2 and A3) and in the inner of slope (A7 and A8), which indicated that the seismic force of inertia exhibited greater strength in the middle/upper sections of slope for the rock mass. Meanwhile, the seismic reflection and superposition commonly occurred surrounding the slope shoulder that the point where the slope's summit intersects with its surface, which belonged to the area where slope instability and failure were prone to occur. In addition, the PGA amplification coefficient of slope along the horizontal and vertical paths gradually decreased with increasing dip angle of the bedding planes of slope (GI_S → MI_S → SI_S), i.e., the "Surface effect" and "Elevation effect" of slope gradually weakened, and the energy transmission of the seismic wave along each path was obviously blocked. This showed that the fragmentation degree (with a significant fissure network of secondary joints due to tension damage) of the rock mass in the middle/upper parts of slope under the action of multiple seismic loads was gradually aggravated with the increasing dip angle of the bedding planes of slope, especially for the SI_S with the bedding planes were not exposed but parallelled to the slope surface.

It can be seen from Fig. 4.7b, d, f that the PGA amplification coefficient of slope along the path of the slope surface (A3 → A2 → A1) and the path of the inner of slope (A8 → A7 → A6) decreased with the increasing seismic action times, and the slope along each vertical path also showed the response characteristics of "Elevation effect". The reduced amplitude of the PGA amplification coefficient of slope was low during the action stage of microseismic waves, but was obvious in both action stages of low and high seismic waves. Specifically, compared to the Stage DA1, the PGA amplification coefficient of slope in the Stage DA2 decreased slightly, i.e., the DCD of slope was not significant under the action of microseismic waves with low amplitude. The PGA amplification coefficient of slope in the Stage DA3 attenuated obviously, i.e., the DCD of the bedding planes and secondary joints of slope was aggravated, and the damage and failure effects of the SW on the slope were more severe than those of the NW. The attenuation amplitude of the PGA amplification coefficient of slope in the Stages DA4 and DA5 was extremely significant (the acceleration response of both the monitoring points A2 and A3 located on the slope surface was relatively approximate), i.e., the DCD and ISF in the inner of slope and the rock mass in the HB were exceedingly severe, and the vertical energy transmission of the seismic wave was basically interrupted. Additionally, due to the significant impact of the fragmentation degree and filtering effect of the middle/upper rock mass of slope, the PGA amplification coefficient of slope along the path of the slope surface and the path of the inner of slope decreased with increasing dip angle of bedding planes of slope. Given the impact of the deterioration of rock mass in the HB and the acceleration response features of slope, generally, the overall instability probability of the MI_S with the bedding planes were exposed on the slope surface and pointed out of slope was greater than that of the GI_S, while that of the SI_S with the bedding planes were not exposed but parallelled to the slope surface was the lowest.

4.4.2 Dynamic Response Rules of the C_D of Slope

Figure 4.8 depicts the variation curves of the C_D of each monitoring point (D1 ~ D4) after the three typical BRSs above-mentioned underwent initiation ~ 500 microseismic waves (Stage CD1) → 500 ~ 1000 microseismic waves (Stage CD2) → epicenter ~ 500 low seismic waves (Stage CD3) → 500 ~ 1000 low seismic waves (Stage CD4) → 100 high seismic waves (Stage CD5).

It can be seen from Fig. 4.8 that the C_D of slope at the monitoring points D1 ~ D4 gradually increased with the action of multiple seismic loads. Specifically, the C_D of slope increased slightly in the Stages CD1 ~ CD3, i.e., the damage of bedding planes and secondary joints of slope was in a state of gradual development at this time. The increased amplitude of the C_D of slope appeared obviously in the Stage CD4, i.e., the cumulative damage effect of slope was more significant at this time, and the slope was constantly approaching the critical instability state. The C_D of slope increased suddenly in the Stage CD5, i.e., the sliding deformation of the rock mass of slope was extremely significant at this time, and the instability and failure occurred in the slope.

It can also be seen from Fig. 4.8 that the C_D of slope along the D1 → D2 → D3 → D4 path increased first, then decreased, and ultimately increased in each action stage of seismic loads. It should be emphasized that the first abrupt increase (at position D2) in the C_D of slope in the above stages was caused by the double superposition of "Surface effect" and "Elevation effect" on the slope shoulder. The second abrupt increase (at position D4) was caused by the transverse through cracks along the junction between the deterioration area and the upper rock mass after the rock mass in the HB was subjected to the combined action of shock, corrosion and scouring.

According to Fig. 4.8, in particular, in each action stage of seismic loads, for the GI_S and MI_S with the bedding planes were exposed on the slope surface and pointed out of slope, the larger the dip angle of bedding planes was, the more significant the C_D of each monitoring point. Although the dip angle of bedding planes of the SI_S was large, the bedding planes were not exposed but parallelled to the slope surface, and thus the C_D of each monitoring point of this type of slope was smaller than that of the above two types of slopes.

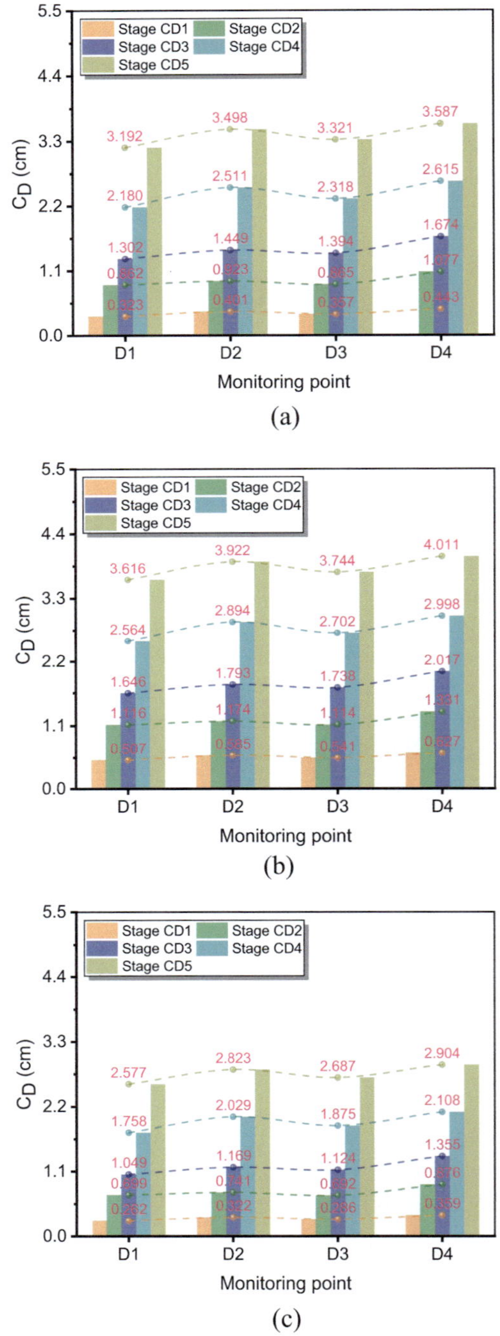

Fig. 4.8 Variation curves of the C_D of slope: **a** GI_S, **b** MI_S, and **c** SI_S; Stages CD1, CD2, CD3, CD4 and CD5 represent "initiation ~ 500 microseismic waves", "500 ~ 1000 microseismic waves", "epicenter ~ 500 low seismic waves", "500 ~ 1000 low seismic waves" and "100 high seismic waves", respectively

4.4.3 Dynamic Response Rules of the R_P of Slope

Figure 4.9 shows the variation curves of the R_P of each monitoring point (R3, R5, R6, R7 and R8) after the three typical BRSs above-mentioned underwent initiation ~ 500 microseismic waves (Stage RP1) → 500 ~ 1000 microseismic waves (Stage RP2) → epicenter ~ 500 low seismic waves (Stage RP3) → 500 ~ 1000 low seismic waves (Stage RP4) → 100 high seismic waves (Stage RP5).

It can be seen from Fig. 4.9 that the R_P at the monitoring points R3, R5, R6, R7 and R8 of slope first increased and subsequently decreased with the action of multiple seismic loads. Specifically, the R_P of slope increased gradually with a small amplitude in the Stages RP1 and RP2, i.e., the DCD and ISF of slope that showed an obvious "Compaction effect" was weak at this time. The R_P of slope first increased and subsequently decreased with a large amplitude in the Stages RP3 and RP4, where the seismic load amplitude increased but did not make the seismic inertia force significantly exceed the critical ISF load of slope, i.e., the energy transmission of the seismic wave was gradually blocked, and the rock mass of slope that approached the critical instability state was gradually broken at this time. The R_P of slope decreased obviously in the Stage RP5, where there was a significant increase in the seismic load amplitude, i.e., the seismic inertia force of slope that showed a significant "Loose effect" was extremely large at this time. Additionally, the R_P of slope along the horizontal path (R3 → R5 → R8) and vertical path (R6 → R7 → R8) in each action stage of seismic loads showed an increasing trend of "Upward-convex arc" and "Downward-concave arc", respectively. In particular, the R_P at each monitoring point in the horizontal and vertical directions of slope generally decreased gradually with increasing dip angle of bedding planes during seismic loading, i.e., the greater the dip angle of bedding planes was, the more severe the fragmentation of the upper rock mass of slope, the faster the contact pressure between rock blocks decreased, and the more obvious the energy transmission of the seismic wave was blocked.

4.4.4 Dynamic Response Rules of the P_W of the HB

Figure 4.10 displays the variation curves of the P_W of each monitoring point (P1 ~ P3) of the HB after the three typical BRSs above-mentioned underwent initiation ~ 500 microseismic waves (Stage PW1) → 500 ~ 1000 microseismic waves (Stage PW2) → epicenter ~ 500 low seismic waves (Stage PW3) → 500 ~ 1000 low seismic waves (Stage PW4) → 100 high seismic waves (Stage PW5).

It can be seen from Fig. 4.10 that the P_W at the monitoring points P1 ~ P3 of the HB increased gradually with the action of multiple seismic loads. Specifically, the increment of the P_W of the HB was relatively small in the Stages PW1 ~ PW3, i.e., there was no obvious secondary cumulative damage fissure in this area, and the water seepage channel was nearly in the closed state at this time. The increment of the P_W of the HB was enhanced in the Stage PW4, i.e., several secondary fractures in this

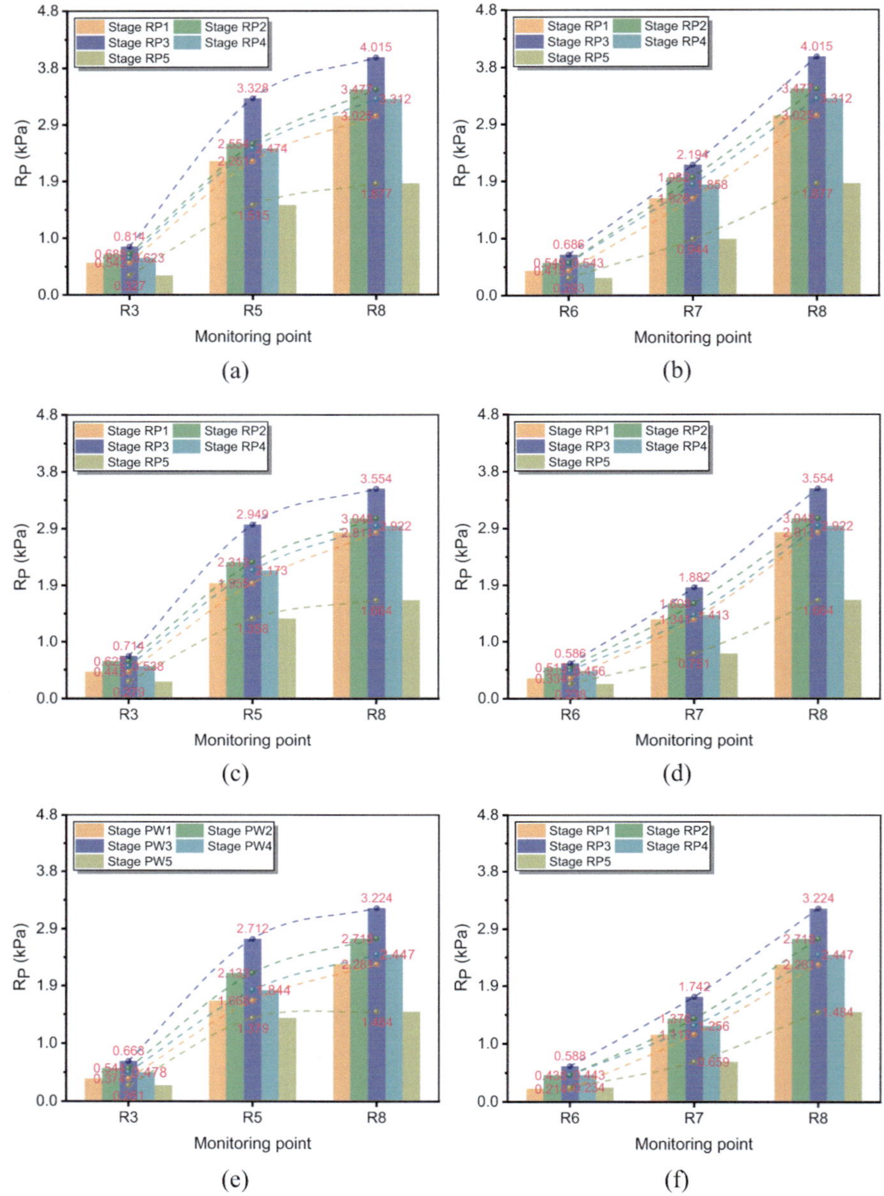

Fig. 4.9 Variation curves of the R_P of slope: **a** GI_S—horizontal direction, **b** GI_S—vertical direction, **c** MI_S—horizontal direction, **d** MI_S—vertical direction, **e** SI_S—horizontal direction, and **f** SI_S—vertical direction; Stages RP1, RP2, RP3, RP4 and RP5 represent "initiation ~ 500 microseismic waves", "500 ~ 1000 microseismic waves", "epicenter ~ 500 low seismic waves", "500 ~ 1000 low seismic waves" and "100 high seismic waves", respectively

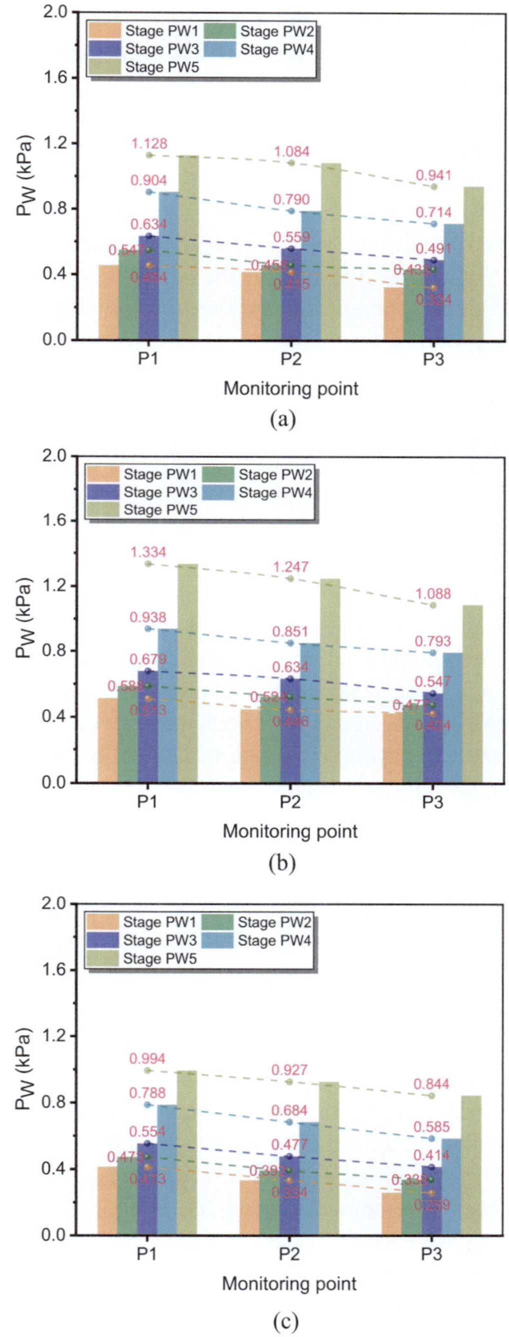

Fig. 4.10 Variation curves of the P_W of the HB: **a** GI_S, **b** MI_S, and **c** SI_S; Stages PW1, PW2, PW3, PW4 and PW5 represent "initiation ~ 500 microseismic waves", "500 ~ 1000 microseismic waves", "epicenter ~ 500 low seismic waves", "500 ~ 1000 low seismic waves" and "100 high seismic waves", respectively

area developed significantly (interwoven distribution of multiple macro cumulative damage fractures), and the water seepage velocity accelerated at this time. The P_W of the HB increased sharply in the Stage PW5, i.e., this area fractured (the rock mass was severely broken) along the upper prefabricated structural planes (joints) following the severe combined action of shock, corrosion and scouring, and the water seepage damage effect was particularly significant at this time. Additionally, the P_W of the HB along the P1 → P2 → P3 path in each action stage of seismic loads exhibited an approximately linearly decreasing trend.

It can also be seen from Fig. 4.10 that the P_W of the HB of the MI_S with the bedding planes were exposed on the slope surface and pointed out of slope was higher than that of the GI_S in each action stage of seismic loads, while that of the SI_S with the bedding planes were not exposed but parallelled to the slope surface was lower than the former two. Specifically, when the bedding planes were exposed on the slope surface and pointed out of slope, the water directly penetrated into the inner of slope along the bedding planes under continuous vibration, which provided favorable conditions for the expansion of the depth, opening and length of the prefabricated structural planes (joints) of the HB. Meanwhile, it was easier to promote the rapid increase in the number of seepage channels in this situation; therefore, the increased amplitude of the P_W in this area was relatively large. Additionally, when the bedding planes were not exposed but parallelled to the slope surface, the water first produced corrosion and scouring actions on the surface rock mass in the HB, thus causing it to spall layer by layer under continuous vibration. Subsequently, the bedding planes of slope were gradually damaged due to the seepage effect after the seepage channels gradually opened from the surface to the inside of the HB; therefore, the P_W response in this area was relatively weak.

4.4.5 Dynamic Response Rules of the D_R and N_F of Slope

The response data of the acceleration time-history at the monitoring points A1 ~ A8 during the whole process of seismic wave loading were obtained after scanning the three typical BRSs above-mentioned using the WN random wave. Based on this, in combination with the relevant theories of transfer functions, the average values of the D_R and N_F of all monitoring points were taken, and then the variation curves of both with the white noise serial number (W_N) were obtained, as shown in Fig. 4.11. The evolution curves of the D_R and N_F with the W_N could be roughly assigned to three development stages, i.e., Stage DN1 (initiation ~ 1000 microseismic waves, i.e., $W_N = 1 \sim 22$), Stage DN2 (epicenter ~ 1000 low seismic waves, i.e., $W_N = 23 \sim 43$) and Stage DN3 (100 high seismic waves, i.e., $W_N = 44 \sim 48$).

It can be seen from Fig. 4.11 that the D_R increased from 11.51% to 15.88% (with an increased amplitude of 4.73%), and the N_F decreased from 26.48 Hz to 21.14 Hz (with a decreased amplitude of 20.17%) for the GI_S under the action of multiple seismic loads. Meanwhile, the D_R increased from 10.45% to 16.27% (with an increased amplitude of 5.82%), and the N_F decreased from 28.12 Hz to 22.12 Hz (with a

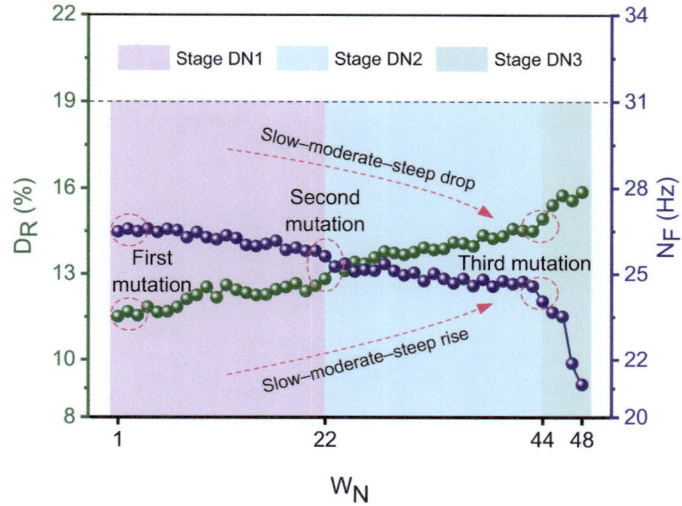

(a)

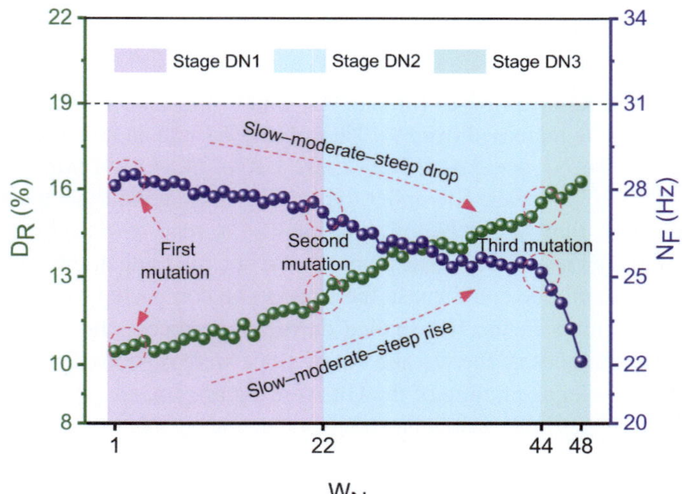

(b)

Fig. 4.11 Variation curves of the D_R (N_F) versus the W_N of slope: **a** GI_S, **b** MI_S, and **c** SI_S; Stages DN1, DN2 and DN3 represent "initiation ~ 1000 microseismic waves, i.e., $W_N = 1 \sim 22$", "epicenter ~ 1000 low seismic waves, i.e., $W_N = 23 \sim 43$" and "100 high seismic waves, i.e., $W_N = 44 \sim 48$", respectively

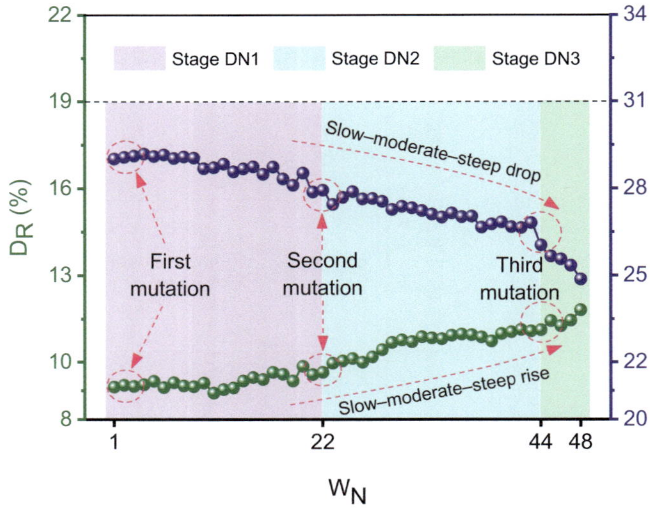

(c)

Fig. 4.11 (continued)

decreased amplitude of 21.34%) for the MI_S under the same loading conditions. Additionally, the D_R increased from 9.13% to 11.81% (with an increased amplitude of 2.68%), and the N_F decreased from 29.02 Hz to 24.87 Hz (with a decreased amplitude of 14.30%) for the SI_S under the same loading conditions. Furthermore, in the Stage DN1, the D_R and N_F of slope showed a trend of slow rise and slight increase first and then slow decrease, respectively, i.e., the finite number of dynamic loads with low amplitude could cause the slope to be compacted by vibration, and the surface rock mass in the HB spalled during this process. In the Stage DN2, the variation amplitudes of the D_R and N_F of slope were obvious, i.e., the fracture development and seepage failure of the HB were significant, and the DCD of slope intensified until it was in the critical instability state during this process. In the Stage DN3, the D_R and N_F of slope exhibited an abrupt rise and steep drop, respectively, i.e., the HB produced overall sliding and the slope was destroyed during this process.

It can also be seen from Fig. 4.11 that the D_R and N_F of slope had three abrupt changes in varying degrees during the seismic action, which reflected the transmission characteristics of the seismic load energy and the damage evolution mechanism of rock mass of slope. For the slope with the bedding planes were exposed on the slope surface and pointed out of slope, the gradient of bedding planes of the MI_S was greater than that of the GI_S, which caused the secondary joints of the MI_S to be more severely cracked (the more broken the rock mass was), and the probability of sliding ISF of the upper rock mass layer by layer along the bedding planes was greater. At this time, the dynamic response features of slope along the horizontal and vertical directions were more significant, i.e., the increased amplitude of the D_R and the decreased amplitude of the N_F of the MI_S were more obvious than those of the GI_S. Additionally, the variation amplitudes of the D_R and N_F of the SI_S with the bedding planes were not exposed but parallelled to the slope surface showed obvious attenuation characteristics compared with the GI_S and MI_S.

4.4.6 Nonlinear Mathematical Model of the DCD of Slope

According to the average value of the N_F at all monitoring points (A1 ~ A8) of slope which were obtained in Sect. 4.4.5 and referring to relevant studies (Nakata and Snieder, 2014; Xu et al., 2022a, c), the variation curves of the cumulative damage degree (C_{DG}) with the W_N of the three typical BRSs above-mentioned under the action of multiple seismic loads were obtained, as shown in Fig. 4.12. The evolution curves of the C_{DG} with the W_N could be roughly assigned to two development stages, i.e., Stage DG1 (initiation ~ 1000 low seismic waves, i.e., $W_N = 1 \sim 43$) and Stage DG2 (100 high seismic waves, i.e., $W_N = 44 \sim 48$).

It can be seen from Fig. 4.12 that in the Stage DG1, the C_{DG} of slope at the prophase had the characteristics of small reduction, i.e., the RM in the inner of slope was compacted by vibration and the rock mass in the HB was almost complete, which belonged to the initial initiation stage of meso cracks. The C_{DG} of slope at the metaphase had an approximately linearly increasing trend with continuously increasing seismic action times, i.e., the local DCD in the inner of slope and the rock mass in the HB gradually became significant, which belonged to the gradual propagation stage of meso cracks. The C_{DG} of slope at the anaphase tended to be gentle owing to the restricted magnitude of the dynamic stress intensity factor at the macro cracks' tip, i.e., the DCD in the inner of slope and the rock mass in the HB was obvious, which belonged to the stable propagation stage of macro cracks. It was not difficult to see that the evolution curve of the C_{DG} with the W_N of slope roughly showed an "S–type" change feature of slight decline, approximate linear increase and slight increase at the prophase, metaphase and anaphase, respectively. Generally, the mathematical model of the nonlinear DCD of slope could be characterized by a cubic function $C_{DG} = F_{p1}{W_N}^3 + F_{p2}{W_N}^2 + F_{p3}W_N + F_{p4}$ where F_{p1}, F_{p2}, F_{p3} and F_{p4} were fitting parameters of a regression analysis in the Stage DG1. Among them, F_{p1}, F_{p2}, F_{p3} and F_{p4} of the GI_S were -5.2120E-6, 3.3575E-4, -0.0018 and -0.0013, respectively; F_{p1}, F_{p2}, F_{p3} and F_{p4} of the MI_S were -6.4395E-6, 4.5586E-4, -0.0033 and -0.0041, respectively; and F_{p1}, F_{p2}, F_{p3} and F_{p4} of the SI_S were -4.6804E-6, 3.1665E-4, -0.0013 and -0.0056, respectively.

It can also be seen from Fig. 4.12 that in the Stage DG2, when the magnitude of the dynamic stress intensity factor at the macro cracks' tip of slope increased to the rock fracture toughness, the spatial geometric size of such cracks expanded rapidly and the number increased rapidly, thus causing severe irreversible DCD to the integrity of the slope body and the HB, and significantly weakening the bearing capacity of rock mass until complete ISF occurred, which belonged to the rapid propagation stage of macro cracks. It was not difficult to see that the evolution curve of the C_{DG} with the W_N of slope roughly showed a "Steep rise type" change feature of rapid

Fig. 4.12 Variation curves of the C_{DG} versus the W_N of slope: **a** GI_S, **b** MI_S, and **c** SI_S; Stages DG1 and DG2 represent "initiation ~ 1000 low seismic waves, i.e., $W_N = 1 \sim 43$" and "100 high seismic waves, i.e., $W_N = 44 \sim 48$", respectively

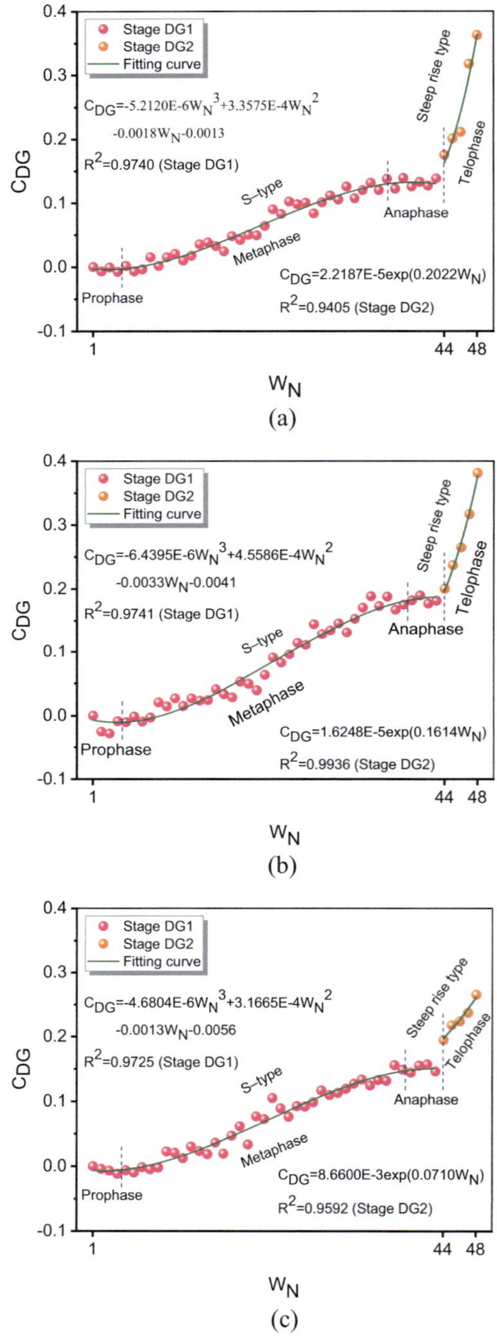

growth. Generally, the mathematical model of the nonlinear DCD of slope could be characterized by an exponential function $C_{DG} = F_{p5}exp(F_{p6}W_N)$ where F_{p5} and F_{p6} were fitting parameters of a regression analysis in the Stage DG2. Among them, F_{p5} and F_{p6} of the GI_S were 2.2187E-5 and 0.2022, respectively; F_{p5} and F_{p6} of the MI_S were 1.6248E-5 and 0.1614, respectively; and F_{p5} and F_{p6} of the SI_S were 8.6600E-3 and 0.0710, respectively.

It can also be seen from Fig. 4.12 that the fitting data of the equations of the nonlinear DCD mathematical models in the Stages DG1 and DG2 for the three typical BRSs above-mentioned were good, which showed that there was an approximate unique relationship between the C_{DG} and the W_N of slope in a certain interval. Additionally, under the action of microseismic waves and low seismic waves, the C_{DG} of the MI_S with the bedding planes were exposed on the slope surface and pointed out of slope was larger than that of the GI_S, while the C_{DG} of the SI_S with the bedding planes were not exposed but parallelled to the slope surface was smaller than the former two. In particular, the evolution curve of the C_{DG} with the W_N was relatively steep in the Stage DG1 (the gradient was especially large at the metaphase) when the DCD and ISF of slope were relatively serious; otherwise, the evolution curve was relatively gentle. The DCD and ISF of slope were rapidly intensified (the abrupt inflection point of this evolution curve was obvious) in the Stage DG2, which generally showed that the final DCD and ISF degree of the consequent slope with the bedding planes were exposed on the slope surface and pointed out of slope was more significant than that of the consequent slope with the bedding planes were not exposed but parallelled to the slope surface.

4.4.7 Evolution Processes of the DCD and ISF of Slope

Figure 4.13 demonstrates the evolution processes of the DCD and ISF of the three typical BRSs above-mentioned under the impact of the deterioration of rock mass in the HB obtained by the model tests. Combined with Figs. 4.11 and 4.12, the evolution processes could be roughly classified as three development stages, i.e., Stage MT1 (initiation ~ 1000 microseismic waves, i.e., $W_N = 1 \sim 22$), Stage MT2 (epicenter ~ 1000 low seismic waves, i.e., $W_N = 23 \sim 43$) and Stage MT3 (100 high seismic waves ~ complete ISF, i.e., $W_N = 44\sim$). It can be seen from Fig. 4.13 that for the GI_S and MI_S with dip angles of 30° and 45°, respectively, these two types of slopes were typical consequent slopes with the bedding planes were exposed on the slope surface and pointed out of slope. Under the above load environment conditions (from Stage MT1 to MT3), the evolution processes of the DCD and ISF had the following characteristics. The upper rock mass of slope had a sliding deformation trend layer by layer along the bedding planes, i.e., the bedding planes were loosened → expanded → penetrated, and the damage degree of secondary joints in the inner of slope was gradually accumulated. Meanwhile, the secondary joints at the trailing edge of slope generally cracked from the slope top and extended and penetrated vertically to the underlying layers in the inner of slope, and the underlying layers often expanded to the top boundary of the HB. In particular, after the formation of vertical tension cracks at the trailing edge of slope, the anti-sliding performance of the middle/upper rock mass of slope was primarily attributed to the shear strength of bedding planes. However, under the repeated action of seismic loads, the bedding planes usually produced the vibration degradation effect of shear strength, thus reducing the shear strength provided by the bedding planes to the rock mass. Therefore, under the combined action of the self-weight of the overlying rock mass and the seismic inertia force, the overall sliding ISF of the middle/upper parts of slope occurred extremely easily. Generally, the rock mass in the middle/upper parts of slope was subject to complete sliding ISF layer by layer along the composite sliding surfaces generated from the convergence of the vertical tension cracks at the rear edge and the through bedding planes in the inner of slope. Meanwhile, each residual sliding surface in the inner of slope was usually smooth, and the vertical residual sliding traces at the rear edge were distributed in a steep step shape. Additionally, under the combined action of shock, corrosion and scouring, the rock mass in the HB gradually went through the evolution processes of the DCD and ISF, such as surface falling, fracture tension, increased seepage, severe fragmentation and detachment sliding. Meanwhile, a significant cavity was formed after detachment and sliding (the deterioration area was transversely connected and fractured along its junction with the upper rock mass), and the overall sliding deformation trend was extremely significant after the upper rock mass gradually lost the support of the lower rock mass. As a whole, after the slope body and the HB underwent ISF, the rock mass densely accumulated at the slope toe in the form of multi-scale broken blocks. The integrity of the rock blocks of the HB was very poor (mostly in small block shape), while the integrity of the upper rock blocks was good (mostly in giant block shape). Meanwhile, the final DCD and ISF degree of rock mass in the HB of the GI_S were weaker than that of the MI_S, and the overall stability of the GI_S was better than that of the MI_S.

It can also be seen from Fig. 4.13 that for the SI_S with a dip angle of 60°, this type of slope was a typical consequent slope with the bedding planes were not exposed but parallelled to the slope surface. Under the above load environment conditions (from Stage MT1 to MT3), the evolution processes of the DCD and ISF had the following characteristics. Under the action of multiple seismic loads, the tension damage cracks first occurred (crack initiation → looseness → slight tension) on the bedding planes near the slope surface at the middle/upper parts of slope, and then most secondary joints within this range were gradually pulled apart due to the bending of bedding planes. At this time, the bedding planes continued to be under tension from the surface to the interior, and the failure surfaces continued to extend, which led to the appearance of a fracture network. The rock mass was in the form of broken blocks and formed by interweaving and converging of the damaged bedding planes and secondary joints, and was densely distributed in the middle/upper parts of slope. Additionally, the fracture network of rock mass gradually densified with increasing seismic action times, and the widths and depths of the fractures expanded simultaneously. At this time, the integrity of rock mass was extremely poor, resulting in layer by layer spalling and sliding ISF of the middle/upper parts

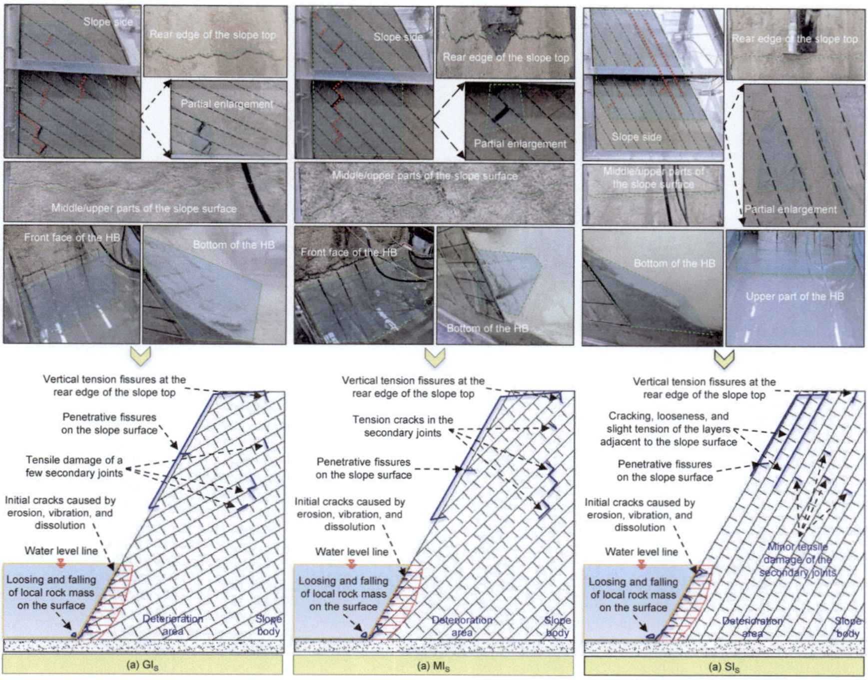

Fig. 4.13 Specified evolution processes of the DCD and ISF of slope: **a** Stage MT1 (initiation ~ 1000 microseismic waves, i.e., $W_N = 1 \sim 22$), **b** Stage MT2 (epicenter ~ 1000 low seismic waves, i.e., $W_N = 23 \sim 43$), and **c** Stage MT3 (100 high seismic waves ~ complete ISF, i.e., $W_N = 44\sim$)

of slope. In particular, in the middle and early action stages of seismic loads, the failure surfaces of the overburden rock mass due to spalling and sliding were mainly generated from the convergence of the steep dip bedding planes at the trailing edge, the step undulating surface in the middle part and the gentle shear surface at the front edge. While in the later action stage of seismic loads, the residual sliding traces on the failure surfaces above-mentioned were finally distributed vertically at the trailing edge in a steep straight shape, and the residual sliding surfaces were smooth. To conclude, the evolution processes of the DCD and ISF of the HB of this type of slope under the combined action of shock, corrosion and scouring were similar to that of the GI_S and MI_S, but the rock mass densely accumulated at the slope toe in the form of multi-scale extremely broken blocks where small blocks accounted for a large proportion after the ISF of the SI_S with the HB. Meanwhile, the final DCD and ISF degree of the HB in the SI_S were weaker than that of the GI_S and MI_S, and the overall stability of the SI_S was better than that of the GI_S and MI_S.

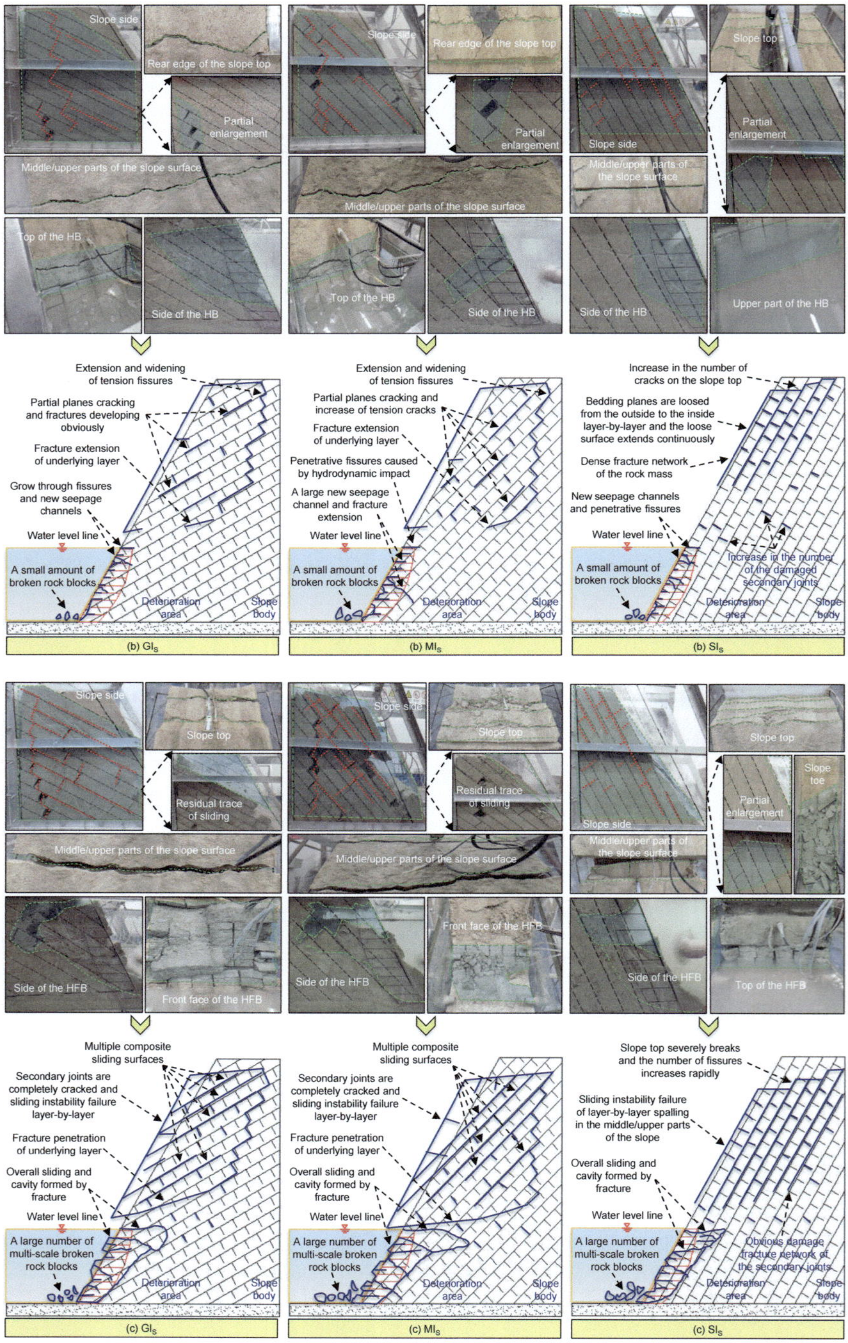

Fig. 4.13 (continued)

4.5 Summary and Conclusions

(1) The deterioration modes of rock mass in the HB could be classified as erosion (denudation) type, corrosion (suffosion) type, avalanche (collapse) type and slip (flow) type. The acceleration responses of the GI_S, MI_S and SI_S showed significant "Surface effect" and "Elevation effect" under seismic loading, and their PGA amplification coefficients constantly weakened. Meanwhile, the C_D, R_P, P_W, D_R and N_F of slope gradually increased, decreased, increased, increased and decreased, respectively. In particular, the order of weakening degree for the PGA amplification coefficient of slope was $SI_S > MI_S > GI_S$. The order of variation amplitude for the C_D, P_W, D_R or N_F of slope was $MI_S > GI_S > SI_S$, while that for the R_P of slope was $GI_S > MI_S > SI_S$.

(2) The mathematical models of the nonlinear DCD of the GI_S, MI_S and SI_S under the action of 2000 microseismic waves (low seismic waves) and 100 high seismic waves could be explained by a cubic function of $C_{DG} = F_{p1}W_N{}^3 + F_{p2}W_N{}^2 + F_{p3}W_N + F_{p4}$ and an exponential function of $C_{DG} = F_{p5}\exp(F_{p6}W_N)$, respectively. The evolution curve of the C_{DG} with the W_N of slope for the former showed an "S–type" change feature of slight decline, approximate linear increase and slight increase at the prophase, metaphase and anaphase, respectively. It showed a "Steep rise type" change feature of rapid growth for the latter. Generally, the C_{DG} of all kinds of slopes increased with continuously increasing seismic action times, and the C_{DG} of the MI_S was larger than that of the GI_S, while the C_{DG} of the SI_S was smaller than that of the former two.

(3) There were three typical stages in the evolution processes of the DCD and ISF for the GI_S and MI_S, i.e., the multiple potential sliding surfaces were formed after the trailing edge of the slope top, secondary joints and bedding planes were cracked, expanded and penetrated; the rock mass slid layer by layer along the composite sliding surfaces generated from the convergence of the trailing edge tension cracks and the penetrated bedding planes in the slope; and the residual sliding traces were distributed in a steep step shape after complete ISF of slope.

(4) There were also three typical stages for the SI_S, i.e., a dense damage fissure network was formed after the trailing edge of the slope top, secondary joints and bedding planes were cracked, expanded and penetrated; the rock mass peeled and slid layer by layer along the failure surface generated from the convergence of the steep dip bedding planes at the trailing edge, the step-like undulating surface in the middle part and the gentle shear surface at the front edge; and the residual sliding traces were distributed in a steep straight shape after complete ISF of slope. In particular, the collapsed rock mass in the HB accumulated at the slope toe in the form of multi-scale broken blocks, and the order of failure degree for the HB was $MI_S > GI_S > SI_S$.

Chapter 5
Sensitivity of Factors Affecting the Dynamic Stability of the BRS Considering Deterioration of Rock Mass in the HB

5.1 Introduction

As the largest water conservancy project in China and even in the whole world, the construction of the TGP has greatly guaranteed the safe water level in the flood season along the Yangtze River, promoted the rapid economic development of the shipping, and eased the tension of power supply in many regions of China (Cheng, 2018; Tang et al., 2019). However, the long and narrow reservoir formed after the completion of the TGP includes over 10 districts and counties (with an area of 1084 km^2) from the reservoir head to the reservoir tail. Its geological environment capacity is extremely limited, and the number of people affected by disasters throughout the year is up to over 20 million (Fourniadis et al., 2007; Zhu, 2010). Especially, the construction of the TGP to some extent interferes with the inherent geological environment balance formed by the long evolvement process of geological history, which makes the reservoir area under complex environment more prone to geological disasters of various types, large-scale and severe damage (Deng et al., 2017; Wang et al., 2019). Among them, the geological disasters such as landslide and collapse of BRS caused by the severe deterioration of rock mass in the HB and the frequent occurrence of the REs since the impoundment are relatively common (Gu & Huang, 2016; Shang et al., 2019), which have seriously endangered the safe operation of shipping in the TGRA and the lives and property of people along the line. Therefore, in order to respond to the national and local major strategic needs and to strengthen the professional construction of geological disaster prevention, it has great theoretical value and practical meaning to conduct research on the dynamic response rules and instability mechanism of the BRS with taking into account the impact of degradation of rock mass in the HB of the TGRA.

Till now, focusing on the degradation mechanism of rock mass in the HB and the dynamic stability of the bank slope in the TGRA, many valuable discussions have been carried out by most scholars, respectively. On the one hand, as the TGRA adopts the water level regulation mode of "storing in winter and discharging in summer",

X. Liu and B. Xu, *Dynamic Stability of the Bedding Rock Slope with Deteriorated Hydro-fluctuation Belt*, https://doi.org/10.1007/978-981-95-6311-1_5

a water level fluctuation belt with an elevation difference of nearly 30 m has been formed along the bank, and the degradation of rock mass in this region has been increasing year by year, which has caused a huge hidden danger to the stability of the bank slope (Gu, 2018). In this regard, Wan et al. (2017) applied rock degradation rules to rock mass, and proved the change features of the failure pattern of the bank slope (with deteriorated HB) in the TGRA with time. Wang et al. (2020) analyzed the stability of large high-steep bank slope in the TGRA via the hydraulic coupling damage model of the HB established by himself. Liang et al. (2021) explored the strength degradation rules of rock mass in the HB of the TGRA via the dry–wet cycle tests and then established the corresponding fracture criteria. Fang et al. (2021) used the seepage tests, joint surface scanning tests and ion concentration measurement tests to explore the seepage characteristics and degradation mechanism of rock mass in the HB of the TGRA. Wang et al. (2017) realized the dynamic operation of the stress–strain constitutive model of dry–wet cycle of rock mass in the fluctuating belt of the TGRA via the secondary development of numerical program. It can be seen that the above studies have revealed the degradation mechanism of rock mass in the HB of the TGRA to some extent, and supported a good foundation for further investigate on the stability of the bank slope. On the other hand, the number of the REs (micro earthquakes) has gradually increased since the impoundment of the TGRA, especially in Badong County, Zigui County and Wushan County along the river and near the shore areas (Li et al., 2019; Tang et al., 2019). Although the seismic wave energy released by a single reservoir earthquake is not easy to cause the direct instability and damage of the bank slope, the bank slope subjected to accumulated action by multiple REs over the years will often produce a series of irresistible sudden secondary geological disasters (Yin et al., 2010; Xu et al., 2022a). In this regard, Ma et al. (2010) has accurately recorded nearly 3000 microseisms and small seisms with magnitude of 0.8 ~ 2.9 in the TGRA and then analyzed their causes of occurrence. Yao et al. (2017b) used the method of stress inversion to predict the earthquake occurrence area and frequency in the TGRA. Jiang et al. (2012), Liu (2017), He (2018), Liu et al. (2018), Yang et al. (2019) and Deng et al. (2020a, b) used the model tests and numerical explorations to systematically investigate the evolvement mechanism of the DCD and ISF of different types of the bank slopes under continuous excitation of multiple microseisms and small seisms. It can be seen that the above studies have enriched the relevant theories of the dynamic stability of the bank slope subjected to microseisms and small seisms in the TGRA to a certain extent, and made up for the shortcomings of most previous studies that only consider strong seism action or a few microseism and small seism actions.

Nevertheless, only a few scholars have discussed the problem of the stability of the bank slope under the coupling effect of the deterioration of rock mass in the HB and the REs in the TGRA at present. Specifically, the UDEC discrete element method, as a commonly used numerical analysis method for discontinuous medium mechanics, treats rock mass as a discontinuous medium composed of rigid or deformed blocks cut from a large number of rock joints. It was often used to simulate the large deformation of rock mass and the process such as rock joint cracking and sliding (Deng et al., 2017; Liu, 2017; He, 2018; Liu et al., 2018c; Yang et al., 2020; Deng et al., 2020a; Xu et al., 2022a, c). Using the UDEC discrete element method to simulate the DCD and ISF process of joint system under quasi-static or dynamic conditions is very suitable, it is an effective method for exploring the dynamic stability of the BRS considering deterioration of rock mass in the HB. To sum up, there are not many reports on the dynamic stability of the bank slope considering the impact of the deterioration of rock mass in the HB of the TGRA, and the sensitivity levels of factors affecting its long-term stability has not been clearly understood. Therefore, it is extremely necessary to conduct in-depth and systematic investigation on this key scientific issue, to provide theoretical support for the analysis and evaluation of the stability of the bank slope in complex environments.

In view of this, based on the research findings of the shaking table model tests, and considering the strength degradation effect of the rock joints, this study conducted a systematic and deep investigation on the sensitivity of factors affecting dynamic stability of the BRS, when considering deterioration of rock mass in the HB. That is to say, according to the results of the UDEC numerical explorations, the response characteristics of the cumulative damage deformation failure, D_{SC} and dynamic sliding velocity of slopes were clarified, and the consistency between the model test findings and numerical exploration data was analyzed and verified by comparison. On this premise, the orthogonal analysis method (OAM) was used to discuss the sensitivity levels of the slope height (S_H), slope angle (S_A), thickness of bedding plane (BP_T), dip angle of bedding plane (BP_{DA}), dynamic load amplitude (DL_A), dynamic load frequency (DL_F), height of water-level of the HB (HB_{HW}), degradation range of the HB (HB_{DR}) and degradation shape of the HB (HB_{DS}) to the impact of the slope's dynamic stability. The research conclusions can provide theoretical basis and reference value for the prevention and treatment of geological disasters of the bank slope, and its long-term stability control.

5.2 UDEC Numerical Exploration Scheme

According to the study of the shaking table model tests, the numerical explorations based on the UDEC, which is an advanced universal computational analysis program developed using the discrete element method as the fundamental theory, were further used to explore the dynamic response mechanism and stability of the three typical BRSs above-mentioned under the impact of the deterioration of rock mass in the HB. Based on the fundamental theory of seepage-stress coupling under dynamic conditions, in combination with the available dynamic stability evaluation methods (mainly including the D_{SC} method based on strength reduction theory and the cumulative permanent displacement method) of the rock slope, and after comparing and analyzing the model test findings and numerical exploration data to verify the reliability of the numerical exploration method in this study, the OAM was used to explore the sensitivity levels of multiple factors such as the slope structural characteristics (S_H, S_A, BP_T and BP_{DA}), dynamic load characteristics (DL_A and DL_F) and water-level (RM deterioration) characteristics of the HB (HB_{HW}, HB_{DR} and HB_{DS}) on the impact of the slope's dynamic stability. Additionally, the HB_{DS} included three types as shown in Fig. 1.3, i.e., Mode A (the DS–1 of erosion (denudation) type), Mode B (the DS–2 of corrosion (suffosion) type) and Mode C (the DS–3 of avalanche (collapse) type).

In particular, the geometric dimensions, grid division, boundary conditions and monitoring point layout of the numerical calculation model are shown in Fig. 5.1 (taking the GI_s as an example, other types of slopes were the same). The physical–mechanical parameters of the rock mass and boundary, rock joints and seepage calculation are demonstrated in Tables 5.1, 5.2 and 5.3, respectively. The numerical working conditions under different influence factors are indicated in Table 5.4.

Among them, the rock mass in the HB was assigned to several irregular discrete small blocks, while the remaining rock mass was assigned to several square discrete small blocks (the corners were adjusted locally); the geometric dimensions of several triangular grid elements were taken as 4.0 m, and the discrete block element used in the numerical exploration was a deformation block. In the initial equilibrium stage, the linear elastic model and the contact-Coulomb slip model were used for the rock

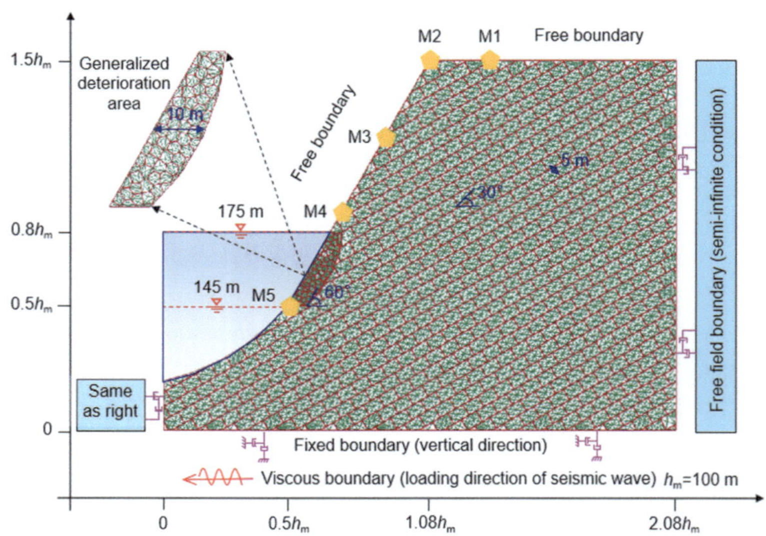

Fig. 5.1 UDEC calculation model of the numerical explorations

Table 5.1 Physical–mechanical parameters of the rock mass and boundary in the numerical explorations

Item	Density (g/cm^3)	Bulk modulus (GPa)	Shear modulus (GPa)	Cohesion (kPa)	Internal friction angle (°)	Tensile strength (kPa)
Rock mass	2.65	4.23	2.66	5260	44.53	526
Boundary	2.65	4.23	2.66	–	–	–

Table 5.2 Physical–mechanical parameters of the rock joints in the numerical explorations

Item	Normal stiffness (GPa/m)	Tangential stiffness (GPa/m)	Cohesion (kPa)	Internal friction angle (°)	Tensile strength (kPa)
Bedding planes of slope	16	12	90	30	60
Secondary joints of slope	12	9.8	8	38	6
Bedding planes of the HB	10.8	8.6	6.4	28	3.6

Table 5.3 Physical–mechanical parameters of the seepage calculation in the numerical explorations

Item	Fluid density (g/cm^3)	Equivalent permeability coefficient (1/Pa·s)	Initial gap width ($\times 10^{-4}$ m)	Residual gap width ($\times 10^{-4}$ m)
Bedding planes of slope	1	83.3	5	2
Secondary joints of slope				
Bedding planes of the HB				

mass and the rock joints, respectively; the Mohr–Coulomb model and the contact-Coulomb slip model were used for the two in the dynamic analysis stage, respectively; and the linear elastic model was used for the free field boundary and the viscous boundary in the dynamic analysis stage. Meanwhile, in the initial equilibrium stage, the top and surface of slope were set as free boundaries, the right and left horizontal directions and the bottom vertical direction were set as fixed boundaries, and the bottom was set as an impermeable boundary. In the dynamic analysis stage, the top and surface of slope were set as free boundaries, the right and left sides were set as free field boundaries, and the bottom vertical and horizontal (loading direction) directions were set as fixed and viscous boundaries, respectively; the free field boundaries can approximately simulate the semi-infinite condition to avoid distortion when the load wave propagates to the right and left boundaries, and the viscous boundary can reduce the reflection superposition when the load wave propagates to the bottom boundary. The bottom was set as an impermeable boundary, and the P_W was applied to the rock mass in the HB as the boundary load conditions.

On the premise of ensuring the calculation requirements and accuracy, and considering the strength degradation effect of the rock joints, the local damping (damping coefficient is 0.15) was chosen for dynamic analysis, and a sinusoidal shear wave with the acceleration time-history equation of $a = \lambda \sin(2\pi ft)$ (λ, f and t are the amplitude, frequency and duration of the loading wave, respectively) was selected to apply

Table 5.4 Numerical working conditions of under different influence factors

Working condition number	S_H (m)	S_A (°)	BP_T (m)	BP_{DA} (°)	DL_A ($\times g$)	DL_F (Hz)	HB_{HW} (m)	HB_{DR} (m)	HB_{DS}
*1	100	60	5	45	0.15	30	175	10	DS–1
*2	80	60	5	45	0.15	30	175	10	DS–1
*3	150	60	5	45	0.15	30	175	10	DS–1
*4	100	75	5	45	0.15	30	175	10	DS–1
*5	100	90	5	45	0.15	30	175	10	DS–1
*6	100	60	3	45	0.15	30	175	10	DS–1
*7	100	60	8	45	0.15	30	175	10	DS–1
*8	100	60	5	30	0.15	30	175	10	DS–1
*9	100	60	5	60	0.15	30	175	10	DS–1
*10	100	60	5	45	0.06	30	175	10	DS–1
*11	100	60	5	45	0.40	30	175	10	DS–1
*12	100	60	5	45	0.15	10	175	10	DS–1
*13	100	60	5	45	0.15	20	175	10	DS–1
*14	100	60	5	45	0.15	30	145	10	DS–1
*15	100	60	5	45	0.15	30	160	10	DS–1
*16	100	60	5	45	0.15	30	175	5	DS–1
*17	100	60	5	45	0.15	30	175	15	DS–1
*18	100	60	5	45	0.15	30	175	10	DS–2
*19	100	60	5	45	0.15	30	175	10	DS–3

seismic loads. Additionally, five C_D monitoring points (Nos. M1 ~ M5) were set at the top, shoulder, surface and toe of slope, the distance between M1 and M2 was 25 m, and the distance between M2 and M3 or between M3 and M4 or between M4 and M5 was approximately 35 m.

5.3 Comparison Between the Numerical and Experimental Results

5.3.1 Dynamic Response Features of the Instability Deformation of Slope

Figure 5.2 depicts the variation curves of the D_{SC} of the three typical BRSs above-mentioned with the seismic action times under the impact of the deterioration of rock mass in the HB.

It can be seen from Fig. 5.2 that the D_{SC} of slope showed a gradually decreasing trend of change under the action of multiple seismic loads, and the evolution curve of the D_{SC} of slope versus the seismic action times could be roughly divided into a slow decline in the early stage and a steep drop in the later stage. Meanwhile, the dynamic instability probability of slope was the MI_S (its S_{SC} was 2.05, D_{SC} of critical instability was 1.04 (approximately 1800 seismic waves), and D_{SC} of ISF was 0.81) > GI_S (its S_{SC} was 2.35, D_{SC} of critical instability was 1.15 (approximately 1800 seismic waves), and D_{SC} of ISF was 0.87) > SI_S (its S_{SC} was 15.85, D_{SC} of critical instability was 2.97 (approximately 1800 seismic waves), and D_{SC} of ISF was 0.98).

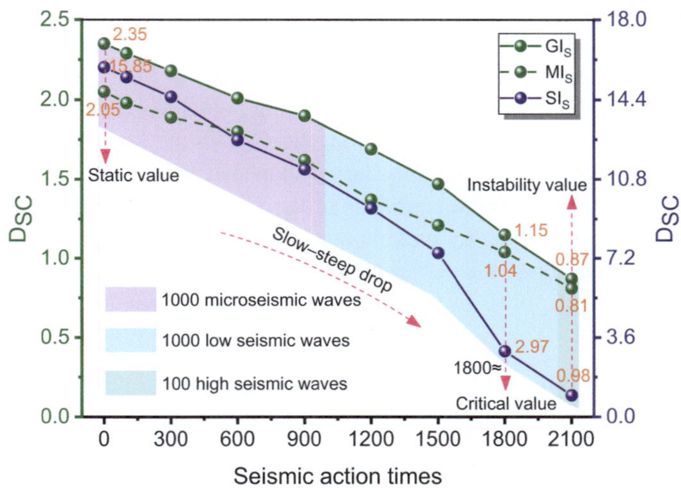

Fig. 5.2 Variation curves of the D_{SC} of slope

This result was closely related to the fact that the bedding planes of the GI_S and MI_S were exposed on the slope surface and pointed out of slope, and the bedding planes of the SI_S were not exposed but parallelled to the slope surface. In particular, the reduced amplitudes of the D_{SC} of the GI_S, MI_S and SI_S after multiple microseismic waves and low seismic waves were 54.06%, 49.27% and 81.26%, respectively; thus, there was an indication that the adverse effect of frequent microseismic waves and low seismic waves on the slope's stability could not be underestimated. In summary, for the response characteristics of the D_{SC} of slope subjected to multiple seismic loads, it can be seen by making a combination with Sect. 4.4.7 that the model test findings and numerical exploration data were basically consistent.

Figure 5.3 shows the variation curves of the horizontal and vertical C_D of each monitoring point of the three typical BRSs above-mentioned with the seismic action times under the impact of the deterioration of rock mass in the HB.

It can be seen from Fig. 5.3 that the horizontal and vertical C_Ds of slope at the top (M1), shoulder (M2), surface (M3 and M4) and toe (M5) of slope show gradually increasing trends, and the incremental exponential function could be used to describe the evolution curve of the C_D of slope versus the seismic action times. In general, the increase in the amplitude of the C_D of slope was slight in the action stage of microseismic waves, while it significantly increased in the action stage of low seismic waves (the seismic action times of critical instability were approximately 1800), and it suddenly increased in the action stage of high seismic waves. When the slope was in the state of ISF, the maximum C_D of the GI_S and MI_S occurred at monitoring point M2, while that of the SI_S existed at the monitoring point M4 (near the top boundary of the HB). Meanwhile, the maximum horizontal (vertical) deformations of the GI_S, MI_S and SI_S were 6.287 m (6.150 m), 9.678 m (9.493 m) and 2.883 m (3.050 m), respectively. In particular, for the GI_S and MI_S, the C_D of each monitoring point is M2 > M3 > M4 > M1 > M5 (the C_D was slightly larger in the horizontal direction than that in the vertical direction), and the damage deformation at the monitoring point M5 was significantly different from that at the other monitoring points. For the SI_S, the C_D of each monitoring point was M4 > M2 > M3 > M1 > M5, except that the C_Ds in the vertical direction of monitoring points M2 and M3 were slightly larger than that in the horizontal direction, the C_Ds in the horizontal direction of other monitoring points were slightly larger than that in the vertical direction. In summary, for the response characteristics of the C_D of slope subjected to multiple seismic loads, it can be seen by making a combination with Sects. 4.4.2 and 4.4.7 that the model test findings and numerical exploration data were basically consistent.

5.3.2 *Modes of the DCD and ISF of Slope*

Figure 5.4 displays the evolution processes of the DCD and ISF of the three typical BRSs above-mentioned under the impact of the deterioration of rock mass in the HB obtained by the numerical explorations. Combined with the results from Sect. 4.4.7, the evolution processes could also be roughly assigned to three development stages,

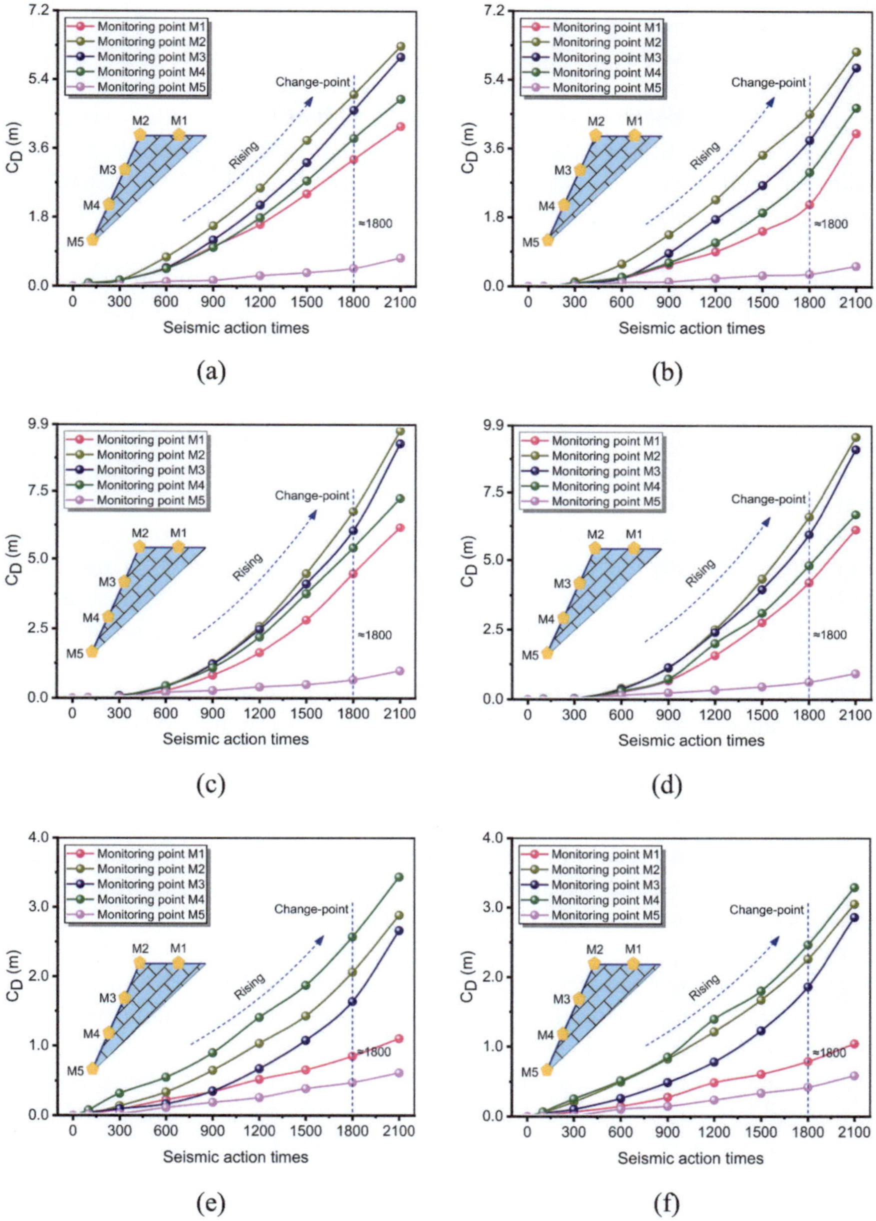

Fig. 5.3 Variation curves of the C_D of slope: **a** GI_S—horizontal direction, **b** GI_S—vertical direction, **c** MI_S—horizontal direction, **d** MI_S—vertical direction, **e** SI_S—horizontal direction, and **f** SI_S—vertical direction

i.e., Stage NE1 (initiation ~ 1000 microseismic waves, i.e., $W_N = 1 \sim 22$), Stage NE2 (epicentre ~ 1000 low seismic waves, i.e., $W_N = 23 \sim 43$) and Stage NE3 (100 high seismic waves ~ complete ISF, i.e., $W_N = 44 \sim$), and comparison showed that the model test findings and numerical exploration data were basically consistent. Specifically, the overall stability of slope in the Stage NE1 was good (stable state), while that in the Stage NE2 was significantly reduced (critical instability state), and that in the Stage NE3 was extremely poor (complete ISF).

According to Fig. 5.4, for the GI_S and MI_S with the bedding planes were exposed on the slope surface and pointed out of slope, the specified characteristics of the DCD and ISF could be summarized as follows. Initially (Stage NE1), the RM in the middle/upper parts of slope underwent slight deformation (the velocity response was generally weak) under the action of multiple microseismic waves (1000 times), i.e., the bedding planes and secondary joints of slope began to sprout microcracks and gradually developed; meanwhile, the rock mass in the HB gradually produced surface looseness and seepage microcracks under the combined action of initial shock, corrosion and scouring. Subsequently (Stage NE2), the overall sliding deformation of the rock mass in the middle/upper parts of slope significantly increased (the velocity response significantly increased) under the action of multiple low seismic waves (1000 times), i.e., the fissures of the bedding planes and secondary joints of slope continued to expand, extend and penetrate; meanwhile, the rock mass in the HB gradually displayed several damage phenomena such as local block falling from the surface to the interior, rapid growth of seepage fissures and accelerated expansion of seepage channels. Finally (Stage NE3), the rock mass in the middle/upper parts of slope underwent complete sliding ISF along multiple composite sliding surfaces (the velocity response suddenly increased) under the action of a few high seismic waves (100 times), i.e., the fracture network formed by the cumulative damage of bedding planes and secondary joints of slope was intertwined and densely distributed; meanwhile, the rock mass in the HB underwent overall sliding failure under the extremely remarkable seepage failure effect, and formed a remarkable detachment cavity.

According to Fig. 5.4, for the SI_S with the bedding planes were not exposed but parallelled to the slope surface, the specified characteristics of the DCD and ISF could be summarized as follows. Initially (Stage NE1), the rock mass near the slope surface in the middle/upper parts of slope slightly deformed with weak a velocity response under the action of multiple microseismic waves (1000 times), i.e., the bedding planes and secondary joints of slope started to crack; meanwhile, the surface rock mass in the HB gradually loosened, and a small amount of seepage channels slightly expanded. Subsequently (Stage NE2), the fragmentation degree of the rock mass in the middle/upper parts of slope increased under the action of multiple low seismic waves (1000 times), i.e., the response velocity and degree of the cumulative damage of bedding planes and secondary joints significantly increased; meanwhile, the rock mass in the HB partially broken, and the number and space size of seepage channels continued to expand (gradually losing the support for the upper sliding mass). Finally (Stage NE3), the rock mass in the middle/upper parts of slope was subject to layer by layer spalling and sliding instability along the failure surfaces under the action of a few high seismic waves (100 times), i.e., the fracture network formed by mutual aggregation

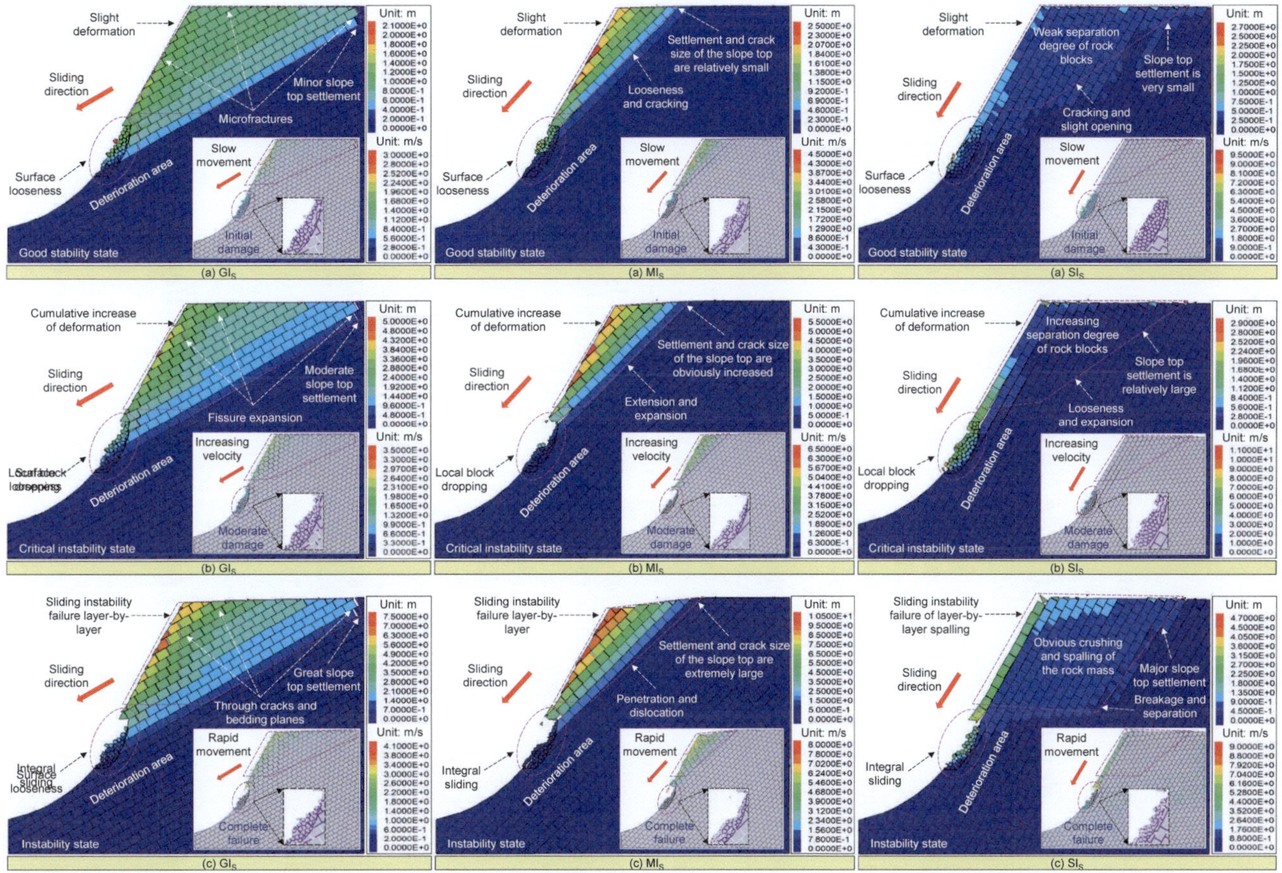

Fig. 5.4 Typical modes of the DCD and ISF of slope: **a** Stage NE1 (initiation ~ 1000 microseismic waves, i.e., $W_N = 1 \sim 22$), **b** Stage NE2 (epicenter ~ 1000 low seismic waves, i.e., $W_N = 23 \sim 43$), and **c** Stage NE3 (100 high seismic waves ~ complete ISF, i.e., $W_N = 44 \sim$)

of the cumulative damage bedding planes and secondary joints of slope were densely distributed; meanwhile, the rock mass in the HB was subject to a significant seepage failure effect, the overall sliding failure occurred, and then a significant detachment cavity formed.

5.4 Factors Influencing the Dynamic Stability of Slope

5.4.1 *Response Rules for the C_D Under Different Influence Factors*

Figure 5.5 shows the variation curves of the C_D of slope under different influence factors (S_H, S_A, BP_T, BP_{DA}, DL_A, DL_F, HB_{HW}, HB_{DR} and HB_{DS}, i.e., working conditions ^{*1} ~ ^{*19}), taking the horizontal and vertical C_D of monitoring point M2 of the slope shoulder with severe dynamic response as an example.

It can be seen from Fig. 5.5 that under the same S_H, S_A, BP_T, BP_{DA}, DL_A, DL_F, HB_{HW}, HB_{DR} or HB_{DS}, both the horizontal and vertical C_D of slope at the all selected monitoring points during the continuous excitation process of multiple seismic loads show a gradual increasing trend, and the evolvement curve of the C_D with the seismic action times can be described in the form of approximately increasing exponential function. That is to say, the slope deformation is relatively minor at the early stage when the seismic action times are small, but it gradually raises with the constant increase of seismic action times (sudden change occurs when approaching the critical instability state), especially the C_D increases suddenly when the slope is completely unstable and damaged. Additionally, the C_D of slope raises with the gradual increase of the S_H, S_A, DL_A, DL_F, HB_{HW} or HB_{DR}, while that decreases with the gradual increases of the BP_T. In particular, the C_D of slope raises with the gradual increases of the BP_{DA} when the bedding planes expose on the slope surface and point out of the slope surface, and the C_D of slope is smaller than the former when the bedding planes do not expose on the slope surface but parallel to the slope surface. The C_D of slope increases with the gradual changes of the HB_{DS} form erosion (denudation) type to corrosion (suffosion) type, and then to avalanche (collapse) type.

5.4.2 *Response Rules for the D_{SC} Under Different Influence Factors*

Figure 5.6 shows the variation curves of the D_{SC} of slope under different influence factors (S_H, S_A, BP_T, BP_{DA}, DL_A, DL_F, HB_{HW}, HB_{DR} and HB_{DS}, i.e., working conditions ^{*1} ~ ^{*19}). It can be seen from Fig. 5.6 that under the same S_H, S_A, BP_T, BP_{DA}, DL_A, DL_F, HB_{HW}, HB_{DR} or HB_{DS}, the D_{SC} of slope tends to decrease slowly first and then to decrease sharply during the continuous excitation process of multiple

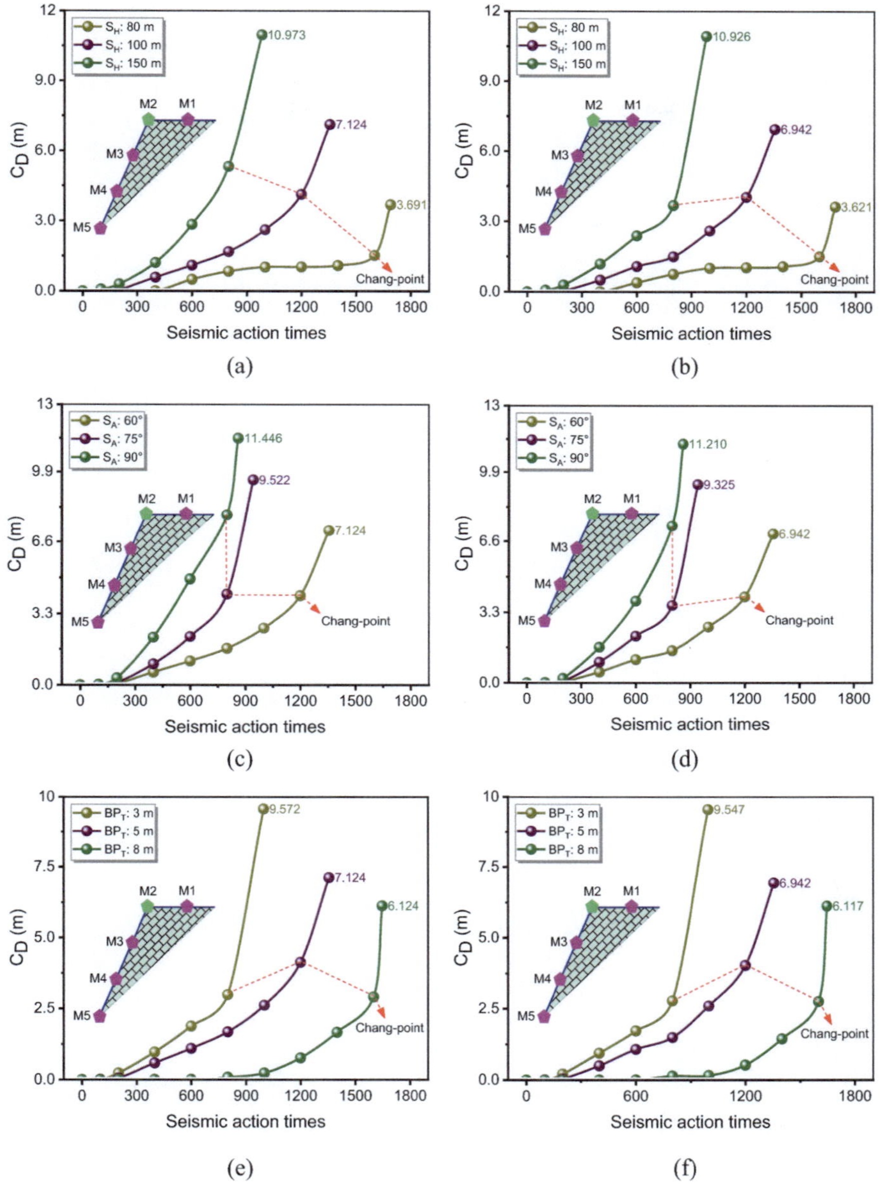

Fig. 5.5 Variation curves of the C_D of slope under different influence factors (working conditions ^{*1} ~ ^{*19}): **a** S_H—horizontal direction, **b** S_H—vertical direction, **c** S_A—horizontal direction, **d** S_A—vertical direction, **e** BP_T—horizontal direction, **f** BP_T—vertical direction, **g** BP_{DA}—horizontal direction, **h** BP_{DA}—vertical direction, **i** DL_A—horizontal direction, **j** DL_A—vertical direction, **k** DL_F—horizontal direction, **l** DL_F—vertical direction, **m** HB_{HW}—horizontal direction, **n** HB_{HW}—vertical direction, **o** HB_{DR}—horizontal direction, **p** HB_{DR}—vertical direction, **q** HB_{DS}—horizontal direction, and **r** HB_{DS}—vertical direction

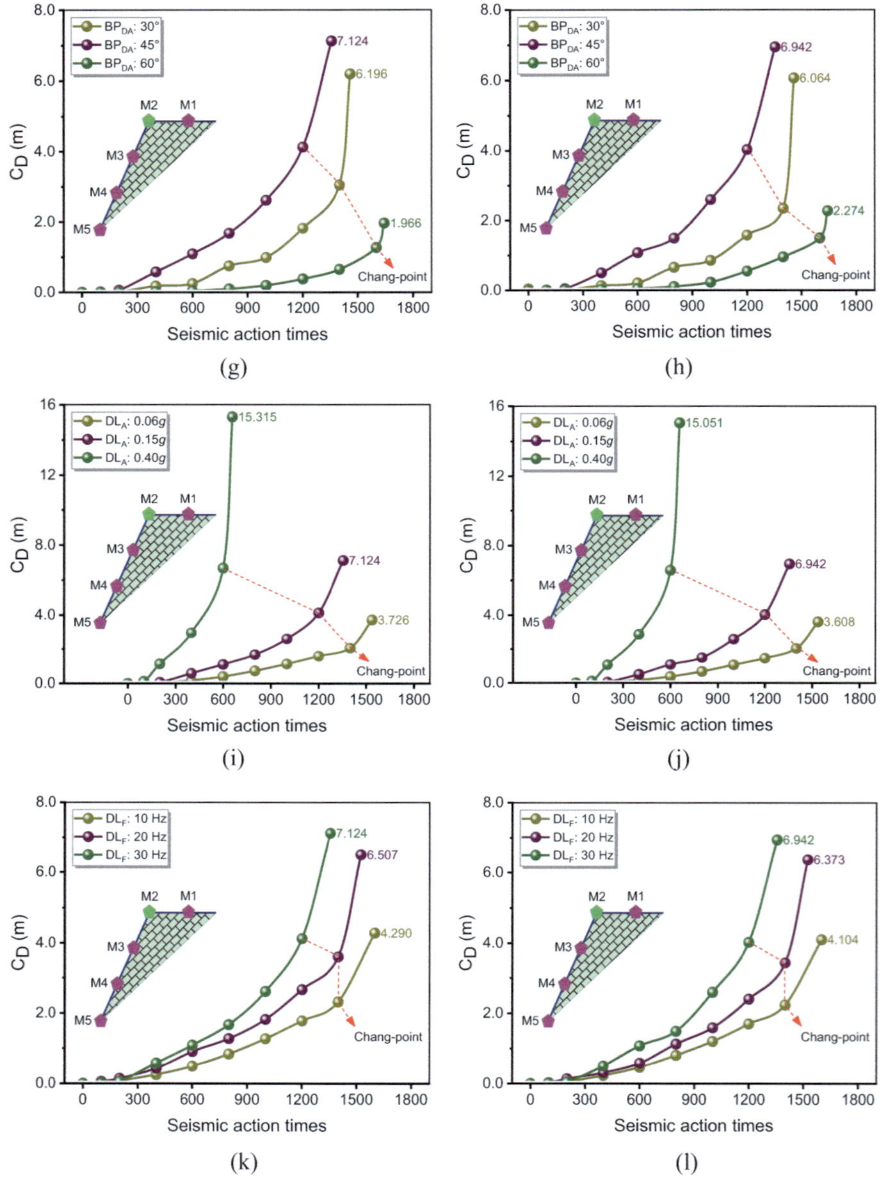

Fig. 5.5 (continued)

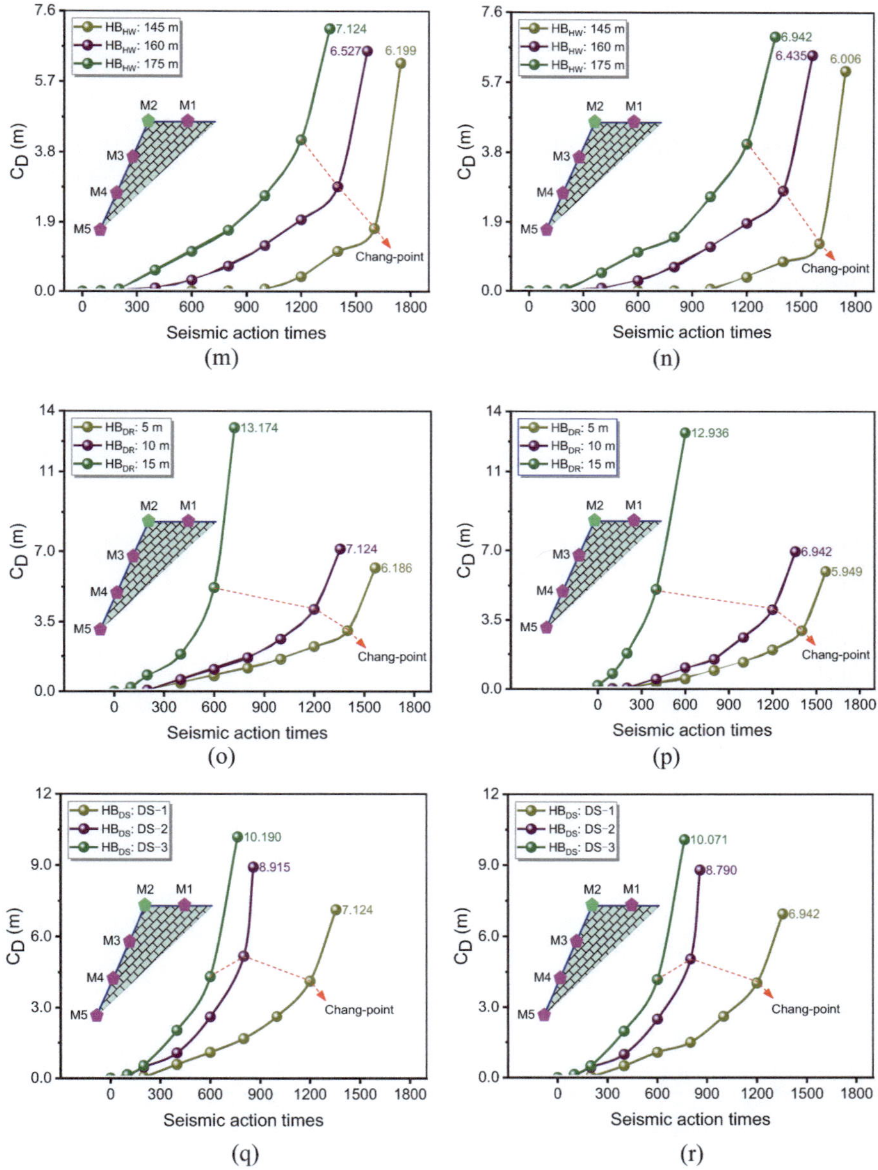

Fig. 5.5 (continued)

seismic loads, and its change range becomes more significant with the continuous increase of seismic action times. Additionally, the D_{SC} of slope drops with the gradual increases of the S_H, S_A, DL_A, DL_F, HB_{HW} or HB_{DR}, while that raises with the gradual increases of the BP_T. In particular, the D_{SC} of slope drops with the gradual increases of the BP_{DA} when the bedding planes expose on the slope surface and point out of

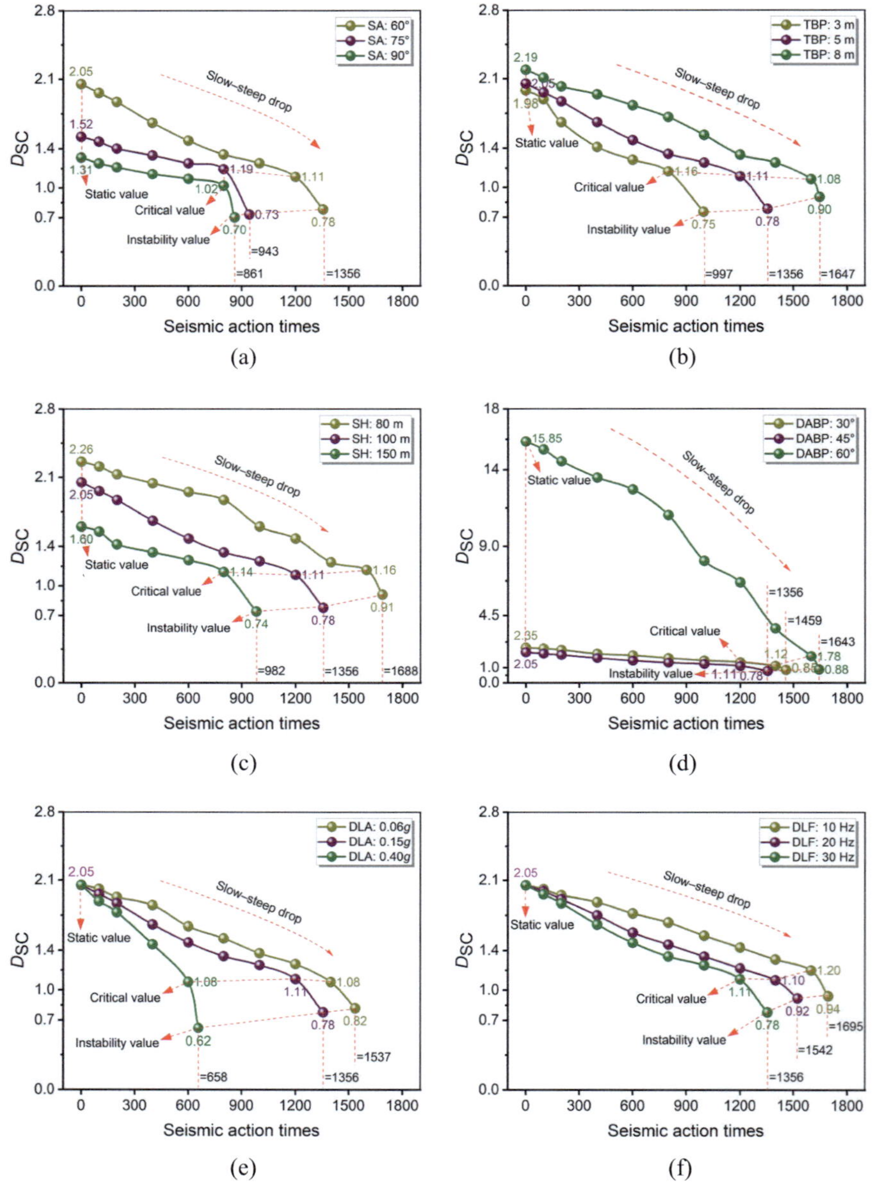

Fig. 5.6 Variation curves of the D_{SC} of slope under different influence factors (working conditions $^{\#}2 \sim {}^{\#}20$): **a** S_H, **b** S_A, **c** BP_T, **d** BP_{DA}, **e** DL_A, **f** DL_F, **g** HB_{HW}, **h** HB_{DR}, and **i** HB_{DS}

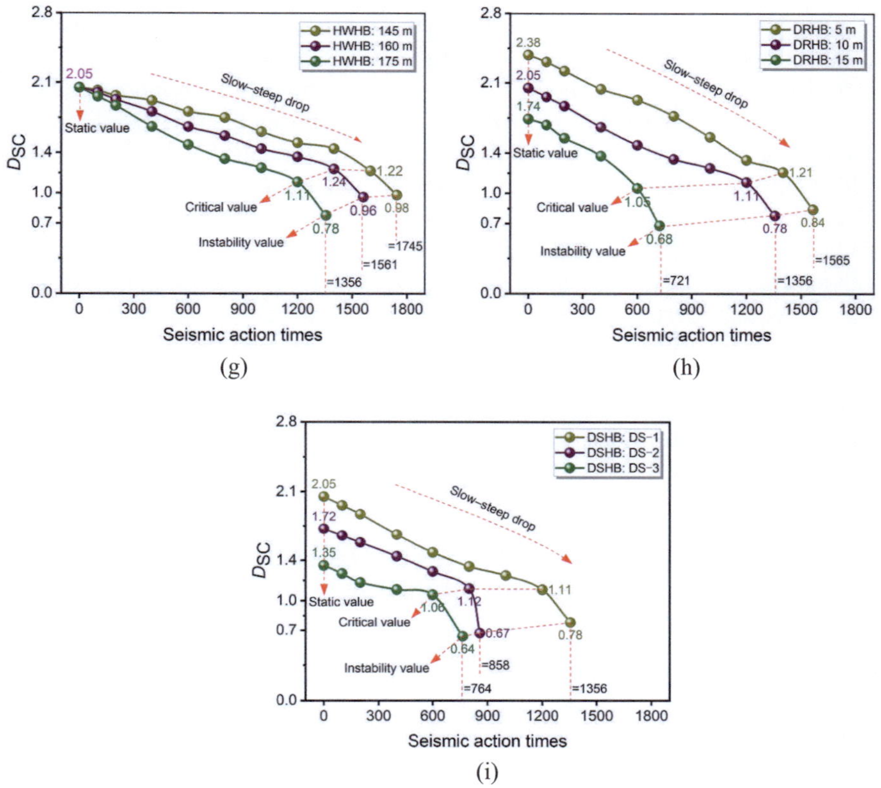

Fig. 5.6 (continued)

the slope surface, and the D_{SC} of slope is larger than the former when the bedding planes do not expose on the slope surface but parallel to the slope surface. The D_{SC} of slope decreases with the gradual changes of the HB_{DS} form erosion (denudation) type to corrosion (suffosion) type, and then to avalanche (collapse) type.

5.4.3 Response Features of Dynamic Instability Under Different Influence Factors

Figure 5.7 shows the variation nephograms of deformation failure and sliding velocity of slope under different influence factors (S_H, S_A, BP_T, BP_{DA}, DL_A, DL_F, HB_{HW}, HB_{DR} and HB_{DS}, i.e., working conditions $^*1 \sim {}^*19$). It can be seen from Fig. 5.7 that the response degree of deformation failure and sliding velocity of slope subjected to continuous excitation of multiple seismic loads shows obvious differences under various influence factors, and this response degree mainly shows that the rock mass in the middle-upper parts of slope and the rock mass in the HB slides and crushes in a progressive process of slight → intensify → remarkable, respectively. That is to say, this response degree affected by the S_A is more significant than the S_H, affected by the BP_{DA} is more significant than the BP_T, affected by the DL_A is more significant than the DL_F, and affected by the HB_{DR} is more significant than the HB_{HW} and the HB_{DS} in turn. In general, the response degree of deformation failure and sliding velocity of slope become more significant with the gradual increases of the S_H, S_A, DL_A, DL_F, HB_{HW} or HB_{DR} under continuous excitation of multiple seismic loads, while that attenuates with the gradual increases of the BP_T. In particular, compared with the condition that the bedding planes expose on the slope surface and point out of the slope surface, the response degree of deformation failure and sliding velocity of slope is weak when the bedding planes do not expose on the slope surface but parallel to the slope surface. This response degree is more significant with the gradual changes of the HB_{DS} form erosion (denudation) type to corrosion (suffosion) type, and then to avalanche (collapse) type.

5.5 Sensitivity Analysis of the Influence Factors on the Dynamic Stability of Slope

5.5.1 Orthogonal Test Design

The consistency between the model test findings and numerical exploration data was analyzed and verified in Sect. 5.3. Meanwhile, the response rules of the D_{SC} and the response characteristics of the dynamic sliding velocity of bedding planes of the three typical BRSs above-mentioned were also discussed in Sect. 5.3, which provided a certain complement to the model test research achievements. To further explore the influence level of S_H, S_A, BP_T, BP_{DA}, DL_A, DL_F, HB_{HW}, HB_{DR} and HB_{DS} on the slope's dynamic stability, the OAM that could objectively reflect the linkage and correlation between various influence factors was used to analyze the sensitivity level of each factor with respect to the impact of the slope's dynamic stability, which could prove the comprehensive impact trends of each factor on the slope's dynamic stability.

In view of this, based on the principle of orthogonal analysis (Schlotfeldt et al., 2018; Ye and Huang, 2020; Xu et al., 2022a, c), three change levels were determined for each of the above nine influence factors. Thus, a total of twenty-seven operating conditions (including four empty columns, as shown in Table 5.5) were calculated and determined according to the orthogonal test table of L_{27} (3^{13}) as shown in Table 5.6. In particular, the D_{SC} and the total C_D (taking the monitoring point M2 of the slope shoulder as an example) in the case of slope instability and failure were selected as the basic evaluation indexes for the sensitivity analysis, and the basic evaluation index values corresponding to various orthogonal operating conditions obtained by the UDEC numerical explorations are shown in Table 5.5.

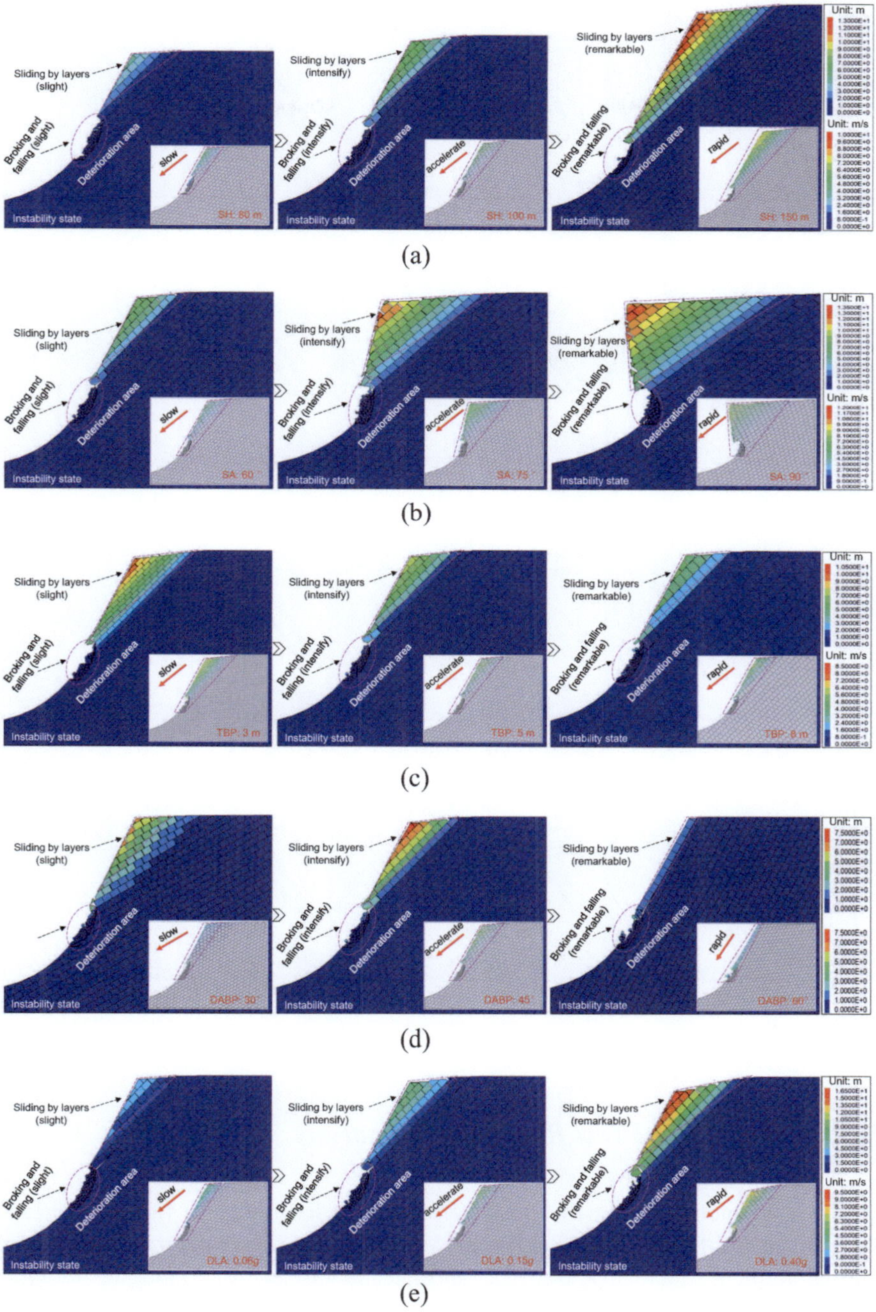

Fig. 5.7 Variation nephograms of deformation failure and sliding velocity of slope under different influence factors (working conditions #2 ~ #20): **a** S_H, **b** S_A, **c** BP_T, **d** BP_{DA}, **e** DL_A, **f** DL_F, **g** HB_{HW}, **h** HB_{DR}, and **i** HB_{DS}

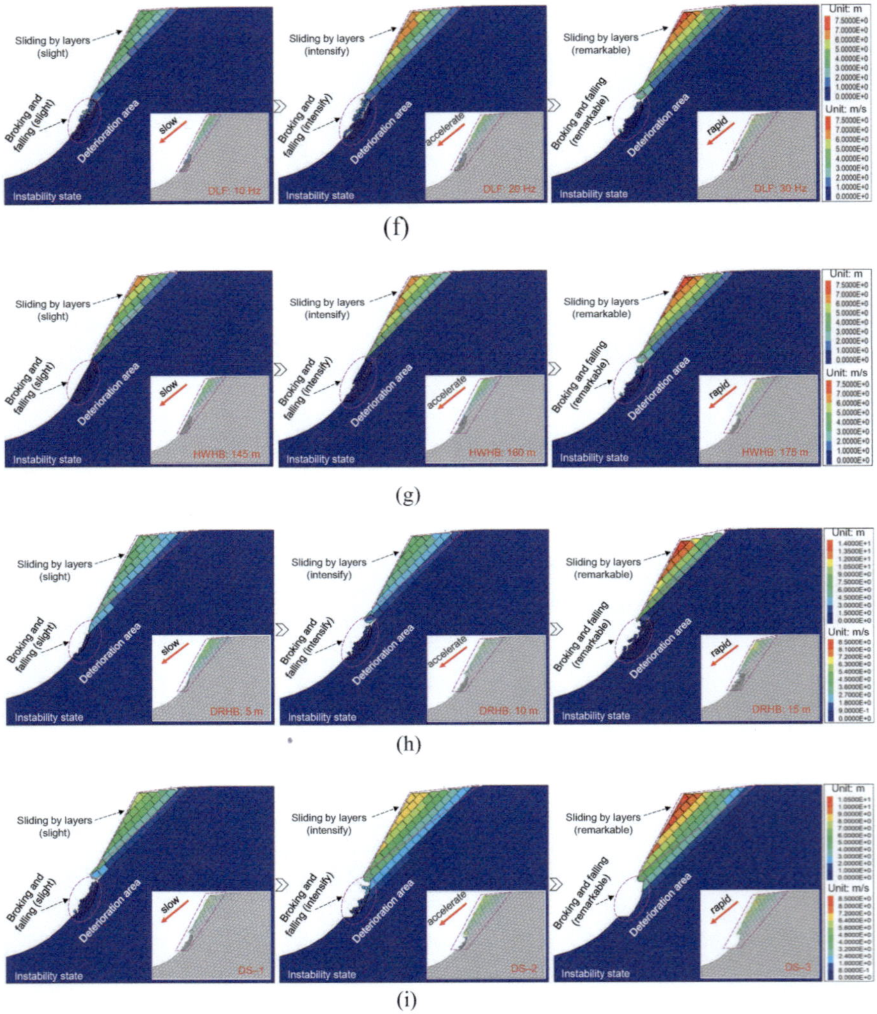

Fig. 5.7 (continued)

5.5.2 *Orthogonal Test Results and Analysis*

Figure 5.8 demonstrates the range analysis results of sensitivity for various influence factors, and Tables 5.7 and 5.8 show the variance analysis results of sensitivity for various influence factors (F indicates an indicator in the F-distribution). It can be seen from Fig. 5.8 and Tables 5.7 and 5.8 that when the significance level $\alpha_x = 0.05$, the DL_A, HB_{DR}, S_A and S_H had highly significant impacts on the slope's stability, and the DL_F and HB_{HW} had moderately significant impacts on the slope's stability, while the HB_{DS}, BP_{DA} and BP_T had impacts of low significance on the slope's stability.

Table 5.5 Specified operating conditions and their basic evaluation index values of the orthogonal tests

Operating condition number	S_H (m)	S_A (°)	BP_T (m)	BP_{DA} (°)	DL_A (× g)	DL_F (Hz)	HB_{HW} (m)	HB_{DR} (m)	HB_{DS}	Empty column				D_{SC}	C_D (m)
#1	1	1	1	1	1	1	1	1	1	1	1	1	1	0.901	8.553
#2	1	1	1	1	2	2	2	2	2	2	2	2	2	0.821	9.528
#3	1	1	1	1	3	3	3	3	3	3	3	3	3	0.885	9.874
#4	1	2	2	2	1	1	1	2	2	2	3	3	3	0.841	8.624
#5	1	2	2	2	2	2	2	3	3	3	1	1	1	0.917	10.011
#6	1	2	2	2	3	3	3	1	1	1	2	2	2	0.931	9.767
#7	1	3	3	3	1	1	1	3	3	3	2	2	2	0.918	9.574
#8	1	3	3	3	2	2	2	1	1	1	3	3	3	0.942	10.125
#9	1	3	3	3	3	3	3	2	2	2	1	1	1	0.904	9.712
#10	2	1	2	3	1	2	3	1	2	3	1	2	3	0.835	9.490
#11	2	1	2	3	2	3	1	2	3	1	2	3	1	0.856	9.481
#12	2	1	2	3	3	1	2	3	1	2	3	1	2	0.949	10.324
#13	2	2	3	1	1	2	3	2	3	1	3	1	2	0.850	9.422
#14	2	2	3	1	2	3	1	3	1	2	1	2	3	0.886	9.970
#15	2	2	3	1	3	1	2	1	2	3	2	3	1	0.876	9.780
#16	2	3	1	2	1	2	3	3	1	2	2	3	1	0.912	10.033
#17	2	3	1	2	2	3	1	1	2	3	3	1	2	0.947	10.839
#18	2	3	1	2	3	1	2	2	3	1	1	2	3	0.898	9.955
#19	3	1	3	2	1	3	2	1	3	2	1	3	2	0.770	10.114
#20	3	1	3	2	2	1	3	2	1	3	2	1	3	0.858	9.127

(continued)

Table 5.5 (continued)

Operating condition number	S_H (m)	S_A (°)	BP_T (m)	BP_{DA} (°)	DL_A (× *g*)	DL_F (Hz)	HB_{HW} (m)	HB_{DR} (m)	HB_{DS}	Empty column				D_{SC}	C_D (m)
#21	3	1	3	2	3	2	1	3	2	1	3	2	1	0.889	9.835
#22	3	2	1	3	1	3	2	2	1	3	3	2	1	0.796	9.580
#23	3	2	1	3	2	1	3	3	2	1	1	3	2	0.900	10.024
#24	3	2	1	3	3	2	1	1	3	2	2	1	3	0.899	10.450
#25	3	3	2	1	1	3	2	3	2	1	2	1	3	0.847	10.031
#26	3	3	2	1	2	1	3	1	3	2	3	2	1	0.911	10.104
#27	3	3	2	1	3	2	1	2	1	3	1	3	2	0.874	10.049

Table 5.6 Influence factors and their levels of the orthogonal tests

Factor	S_H (m)	S_A (°)	BP_T (m)	BP_{DA} (°)	DL_A (× g)	DL_F (Hz)	HB_{HW} (m)	HB_{DR} (m)	HB_{DS}
Level 1	80	60	3	30	0.06	10	145	5	Mode A
Level 2	100	75	5	45	0.15	20	160	10	Mode B
Level 3	150	90	8	60	0.40	30	175	15	Mode C

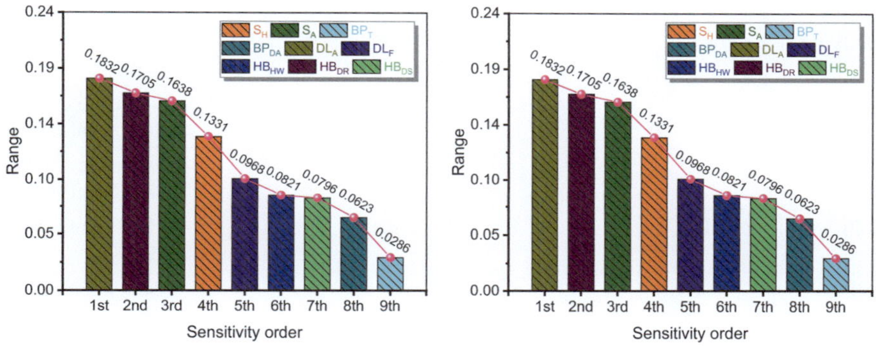

(a) (b)

Fig. 5.8 Range analysis results: **a** Using the index of the D_{SC} of slope, and **b** Using the index of the C_D of slope

Generally, the sensitivity level of each factor to the impact of the slope's stability was in the order of DL_A (1st) > HB_{DR} (2nd) > S_A (3rd) > S_H (4th) > DL_F (5th) > HB_{HW} (6th) > HB_{DS} (7th) > BP_{DA} (8th) > BP_T (9th), and the data generated using the range analysis approach and the variance analysis approach were basically consistent. Therefore, when analyzing the dynamic stability of the bank slopes, the degree of influence of multiple factors such as the slope structural characteristics (S_H, S_A, BP_T and BP_{DA}), dynamic load characteristics (DL_A and DL_F) and water-level (deterioration of rock mass) characteristics of the HB (HB_{HW}, HB_{DR} and HB_{DS}) should be distinguished according to the sensitivity order. This can lay a reliable theoretical foundation for the long-term stability evaluation of the bank slopes, and especially offer reasonable suggestions and measures for the prediction and prevention of geological disasters that often occur in the BRSs of the TGRA.

Table 5.7 Variance analysis results using the index of the D_{SC} of slope

Factor	Sum of deviation squares	Free degree	*F* value	*F* critical value	Ranking
S_H (m)	0.006	2	4.799	$F_{0.05}(2, 8) =$ 4.460	High significant (4th)
S_A (°)	0.009		6.525		High significant (3rd)
BP_T (m)	0.000		0.250		Low significant (9th)
BP_{DA} (°)	0.001		0.993		Low significant (8th)
DL_A ($\times g$)	0.012		9.148		High significant (1st)
DL_F (Hz)	0.003		2.206		Moderate significant (5th)
HB_{HW} (m)	0.003		1.877		Moderate significant (6th)
HB_{DR} (m)	0.010		7.529		High significant (2nd)
HB_{DS}	0.002		1.631		Low significant (7th)

Table 5.8 Variance analysis results using the index of the C_D of slope

Factor	Sum of deviation squares	Free degree	*F* value	*F* critical value	Ranking
S_H (m)	0.926	2	5.730	$F_{0.05}(2, 8) =$ 4.460	High significant (4th)
S_A (°)	0.973		6.021		High significant (3rd)
BP_T (m)	0.087		0.538		Low significant (9th)
BP_{DA} (°)	0.122		0.755		Low significant (8th)
DL_A ($\times g$)	1.235		7.640		High significant (1st)
DL_F (Hz)	0.718		4.439		Moderate significant (5th)
HB_{HW} (m)	0.293		1.815		Moderate significant (6th)
HB_{DR} (m)	1.180		7.297		High significant (2nd)
HB_{DS}	0.129		0.801		Low significant (7th)

5.6 Summary and Conclusions

(1) Although there is an inevitable difference between the model test findings and numerical exploration data to a certain extent, both indicate that the three typical slope models above-mentioned have obvious stage development characteristics of DCD and ISF under multiple continuous seismic loads, and the two results are basically consistent. The C_D of the GI_S, MI_S and SI_S increased in the form of an approximately exponential function, the D_{SC} of that showed a trend of first a slow decline and then a steep drop, and their variation amplitude increased with continuously increasing seismic action times. Fully considering the variation characteristics of the deformation failure and velocity response of the bedding planes, the secondary joints and the rock mass in the HB, the overall stability of the MI_S was less than that of the GI_S, while the overall stability of the SI_S was better than that of the former two.

(2) With the gradual changes of the HB_{DS} form erosion (denudation) type to corrosion (suffosion) type, and then to avalanche (collapse) type, the C_D and D_{SC} of slope increases and decreases, respectively, while the response degree of deformation failure and sliding velocity of slope is more significant with that. This response degree above-mentioned affected by the S_A is more significant than the S_H, affected by the BP_{DA} is more significant than the BP_T, affected by the DL_A is more significant than the DL_F, and affected by the HB_{DR} is more significant than the HB_{HW} and the HB_{DS} in turn. Generally, under the effect of different factors, the C_D of slope subjected to continuous excitation of multiple seismic loads increases in the form of an approximately exponential function, the D_{SC} of slope decreases slowly first and then drops sharply, and the response degree of slope deformation and sliding velocity is obviously different.

(3) Based on the basic principle of the orthogonal test design, the C_D and D_{SC} when the ISF of slope were selected as the basic evaluation indicators for sensitivity analysis, and the sensitivity of the overall impact of various factors on the dynamic stability of slope was obtained through the range analysis approach and the variance analysis approach. The sensitivity level of the influence of different factors on the dynamic stability of slope is in the order of $DL_A > HB_{DR} > S_A > S_H > DL_F > HB_{HW} > HB_{DS} > BP_{DA} > BP_T$ according to the OAM, and it is very necessary to judge the influence level of different factors on the stability of the bank slope in combination with the order of sensitivity when analysing and estimating the stability of slope engineering.

References

Asadi, M. S., Rasouli, V., & Barla, G. (2012). A bonded particle model simulation of shear strength and asperity degradation for rough rock fractures. *Rock Mechanics and Rock Engineering, 45*(5), 649–675.

Bahaaddini, M., Sharrock, G., & Hebblewhite, B. K. (2013). Numerical direct shear tests to model the shear behaviour of rock joints. *Computers and Geotechnics, 51*, 101–115.

Bandis, S. C., Lumsden, A. C., & Barton, N. R. (1983). Fundamentals of rock joint deformation. *International Journal of Rock Mechanics and Mining Sciences & Geomechanics Abstracts, 20*(6), 249–268.

Barton, N., & Choubey, V. (1977). The shear strength of rock joints in theory and practice. *Rock Mechanics and Rock Engineering, 10*(1), 1–54.

Barton, N. (1973). Review of a new shear strength criterion for rock joints. *Engineering Geology, 7*, 287–332.

Cai, M. F., He, M. C., & Liu, D. Y. (2002). *Rock mechanics and engineering*. Science Press.

Chen, J. H., Wei, G. C., Shen, X. L., Dao, M., & Zhao, L. Y. (2010). Introduction on earthquake monitoring in Three Gorges Area before and after impoundment of Three Gorges Reservoir. *Journal of Yangtze River Scientific Research Institute, 27*(5), 85–87. (in Chinese).

Chen, L. Y., & Talwani, P. (1998). Reservoir-induced seismicity in China. *Pure and Applied Geophysics, 153*(1), 133–149.

Cheng, X. (2018). *Research on the sustainable development mechanism in the Three Gorges reservoir region based on poverty inducing model of eco-environment and geo-disaster*. PhD thesis. China University of Geosciences (in Chinese).

Cheng, Y. M., Lansivaara, T., & Wei, W. B. (2007). Two-dimensional slope stability analysis by limit equilibrium and strength reduction methods. *Computers and Geotechnics, 34*(3), 137–150.

Choudhury, D., & Nimbalkar, S. S. (2006). Pseudo-dynamic approach of seismic active earth pressure behind retaining wall. *Geotechnical and Geological Engineering, 24*(5), 1103–1113.

Crawford, A. W., & Curran, J. H. (1981). The influence of shear velocity on the frictional resistance of rock discontinuities. *International Journal of Rock Mechanics and Mining Sciences & Geomechanics Abstracts, 18*(6), 505–515.

Deng, G. D., Li, X. C., & Chen, J. X. (2022). Research on coupling coordination and the development of green shipping and economic growth in China. *Sustainability-Basel, 13*(24), 13901.

Deng, H. F. (2010). *Study on the mechanism and effects of water-rock interaction under water level change region*. PhD thesis. Wuhan University (in Chinese).

Deng, H. F., Yuan, X. F., Wang, L. H., Zhu, M., Luo, Q., & He, M. (2013). Experimental research on changes in the mechanical property law of reservoir bank sandstone under "immersion–air dry" circulation. *Environmental Engineering and Management Journal, 12*(9), 1785–1789.

X. Liu and B. Xu, *Dynamic Stability of the Bedding Rock Slope with Deteriorated Hydro-fluctuation Belt*, https://doi.org/10.1007/978-981-95-6311-1

Deng, Q. L., Fu, M., Ren, X. W., Liu, F. Z., & Tang, H. M. (2017). Precedent long-term gravitational deformation of large scale landslides in the Three Gorges reservoir area. *China. Eng. Geol., 221*, 170–183.

Deng, Z. Y., Liu, X. R., Liu, Y. Q., Liu, S. L., Han, Y. F., Liu, J. H., & Tu, Y. L. (2020a). Model test and numerical simulation on the dynamic stability of the bedding rock slope under frequent microseisms. *Earthquake Engineering and Engineering Vibration, 19*(4), 919–935.

Deng, Z. Y., Liu, X. R., Liu, Y. Q., Liu, S. L., Han, Y. F., Tu, Y. L., & Gu, J. (2020b). Cumulative damage evolution and failure modes of the bedding rock slope under frequent microseisms. *Arabian Journal of Geosciences, 13*(10), 384.

Du, W. J., Sheng, Q., Fu, X. D., Chen, J., & Zhou, Y. Q. (2022). A TPDP-MPM-based approach to understanding the evolution mechanism of landslide-induced disaster chain. *Journal of Rock Mechanics and Geotechnical Engineering, 14*(4), 1200–1209.

Erarslan, N., & Williams, D. J. (2012). Mechanism of rock fatigue damage in terms of fracturing modes. *International Journal of Fatigue, 43*, 76–89.

Fang, J. C., Deng, H. F., Li, J. L., & Assefa, E. (2021). Study on the seepage characteristics and degradation mechanism of a single-jointed sandstone under the cyclic dry-wet process in the Three Gorges reservoir. *Bulletin of Engineering Geology and the Environment, 80*(10), 8123–8136.

Fathi, A., Moradian, Z., Rivard, P., & Ballivy, G. (2016). Shear mechanism of rock joints under pre-peak cyclic loading condition. *International Journal of Rock Mechanics and Mining Sciences, 83*, 197–210.

Feng, P., Dai, F., Liu, Y., Xu, N., & Zhao, T. (2018). Effects of strain rate on the mechanical and fracturing behaviors of rock-like specimens containing two unparallel fssures under uniaxial compression. *Soil Dynamics and Earthquake Engineering, 110*, 195–211.

Fourniadis, I. G., Liu, J. G., & Mason, P. J. (2007). Regional assessment of landslide impact in the Three Gorges area, China, using ASTER data: Wushan-Zigui. *Landslides, 4*(3), 267–278.

Fu, B. J., Wu, B. F., Lu, Y. H., Xu, Z. H., Cao, J. H., Niu, D., Yang, G. S., & Zhou, Y. M. (2010). Three Gorges project: Efforts and challenges for the environment. *Progress in Physical Geography, 34*(6), 741–754.

Fu, Y. (2010). *Study on water-rock interaction with the cyclic drying-wetting effect on rock*. PhD thesis. Chongqing University (in Chinese).

Ge, X. R. (1979). The mechanical behaviors and analog analytical method of joints and weak intercal ations in rock mass. *Rock Soil Mechanics, 1*(1), 54–75. (in Chinese).

Ghosh, P. (2007). Seismic passive earth pressure behind non-vertical retaining wall using pseudo-dynamic analysis. *Geotechnical and Geological Engineering, 25*(5), 693–703.

Gine, J. (2011). On the origin of the inertial force and gravitation. *International Journal of Theoretical Physics, 50*(2), 607–617.

Goodman, R. E. (1984). *Introduction to rock mechanics*. Wiley and Sons.

Gu, D. M. (2018). *Failure mechanism and evolution of the anaclinal high carbonatite rock slopes with soft toes in the Three Gorges Region*. PhD thesis. Chongqing University (in Chinese).

Gu, D. M., & Huang, D. (2016). Complex rock topple-rock slide failure of an anaclinal rock slope in the Wu Gorge, Yangtze River, China. *Engineering Geology, 208*, 165–180.

Gu, D. M., Huang, D., Yang, W. D., Zhu, J. L., & Fu, G. Y. (2017). Understanding the triggering mechanism and possible kinematic evolution of a reactivated landslide in the Three Gorges Reservoir. *Landslides, 14*(6), 2073–2087.

Guo, Y., Wan, W., Zhang, H., & Tang, X. G. (2012). A study of reservoir-induced earthquakes–Taking the Three Gorges Reservoir area for example. *Recent Developments in World Seismology, 4*, 13–18. (in Chinese).

Guo, Y., & Yin, W. Y. (2021). Study on the development situation of disaster discipline in china from the perspective of periodical literature. *Journal of Catastrophe, 36*(4), 126–132. (in Chinese).

He, C. M. (2018). *Mechanical behavior of muddy soft rock and dynamic stability of soft-hard interbedded slopes in reservoir area under cyclic loading*. PhD thesis. Chongqing University (in Chinese).

He, K. Q., & Wang, S. J. (2006). Double-parameter threshold and its formation mechanism of the colluvial landslide: Xintan landslide. *China Environmental Geology, 49*(5), 696–707.

Hoek, E. (1983). Strength of jointed rock masses. *Geotechnique, 33*(3), 187–223.

Hoek, E., & Bray, J. (1981). *Rock slope engineering* (3rd ed.). Institution of Mining and Metallurgy.

Homand, F., Belem, T., & Souley, M. (2001). Friction and degradation of rock joint surfaces under shear loads. *International Journal for Numerical and Analytical Methods in Geomechanics, 25*(10), 973–999.

Huang, B. L., Chen, L. D., Peng, X. M., Liu, G. N., Chen, X. T., Dong, H. G., & Lei, T. C. (2010). Assessment of the risk of rockfalls in Wu Gorge, Three Gorges, China. *Landslides, 7*(1), 1–11.

Huang, B. L., Yin, Y. P., & Tan, J. M. (2019a). Risk assessment for landslide induced impulse waves in the Three Gorges Reservoir, China. *Landslides, 16*(3), 585–596.

Huang, B. L., Yin, Y. P., Yan, G. Q., Li, B., Qin, Z., & Wang, J. (2021). A study on in situ measurements of carbonate rock mass degradation in the water-level fluctuation zone of the Three Gorges Reservoir, China. *Bulletin of Engineering Geology and the Environment, 80*(2), 1091–1101.

Huang, B. L., Yin, Y. P., Zhang, Z. H., Wang, J., Qin, S., & Yan, G. Q. (2019b). Study on deterioration characteristics of shallow rock mass in water the level fluctuation zone of karst bank slopes in Three Gorges Reservoir area. *Chinese Journal of Rock Mechanics and Engineering, 38*(9), 1786–1796. (in Chinese).

Huang, B. L., Yin, Y. P., Li, B., Qin, Z., & Zhang, P. (2020). Rock mass deterioration and its catastrophic effect of karst bank slope in the Three Gorges Project Reservoir area. *Hydrogeology & Engineering Geology, 47*(4), 51–61. (in Chinese).

Huang, R., Zhu, L. P., Encarnacion, J., Xu, Y. X., Tang, C. C., Luo, S., & Jiang, X. H. (2018). Seismic and geologic evidence of water-induced earthquakes in the Three Gorges Reservoir region of China. *Geophysical Research Letters, 45*(12), 5929–5936.

Jafari, M. K., Hosseini, K. A., Pellet, F., Boulon, M., & Buzzi, O. (2003). Evaluation of shear strength of rock joints subjected to cyclic loading. *Soil Dynamics and Earthquake Engineering, 23*(7), 619–630.

Jian, W. X., Xu, Q., Yang, H. F., & Wang, F. W. (2014). Mechanism and failure process of Qianjiangping landslide in the Three Gorges Reservoir, China. *Environmental Earth Sciences, 72*(8), 2999–3013.

Jiang, J. W. (2012) *Research on the deformation mechanism and dynamic response of typical landslides in Three Gorges Reservoir in case of frequent microseisms.* PhD thesis. China University of Geosciences (in Chinese).

Jiang, J. W., Xiang, W., Zhang, W., & Pan, J. J. (2016). Deformation forecasting of Huangtupo riverside landslide in the case of frequent microseisms. *Journal of Earth Science, 27*(1), 160–166.

Kang, J. Q., Zhu, J. B., & Zhao, J. (2019). A review of mechanisms of induced earthquakes: From a view of rock mechanics. *Geomechanics and Geophysics for Geo-Energy and Geo-Resources, 5*(2), 171–196.

Kasama, K., Furukawa, Z., & Hu, L. H. (2021). Practical reliability analysis for earthquake-induced 3D landslide using stochastic response surface method. *Computers and Geotechnics, 137*, 104303.

Kou, M. M., Liu, X. R., Tang, S. D., & Wang, Y. T. (2019). Experimental study of the prepeak cyclic shear mechanical behaviors of artificial rock joints with multiscale asperities. *Soil Dynamics and Earthquake Engineering, 120*, 58–74.

Lambert, C., & Coll, C. (2014). Discrete modeling of rock joints with a smooth-joint contact model. *Journal of Rock Mechanics and Geotechnical Engineering, 6*(1), 1–12.

Lee, H. S., Park, Y. J., Cho, T. F., & You, K. H. (2001). Influence of asperity degradation on the mechanical behaviors of rough rock joints under cyclic shear loading. *International Journal of Rock Mechanics and Mining Sciences, 38*(7), 967–980.

Li, H. B., Jiang, H. J., Zhao, J., Huang, L. X., & Li, J. R. (2003). Some problems about safty analysis of rock engineering under dynamic load. *Chinese Journal of Rock Mechanics and Engineering, 22*(11), 1887–1891. (in Chinese).

Li, C., Su, L. J., Liao, H. J., Zhang, C. L., & Xiao, S. Y. (2021a). Modeling of rapid evaluation for seismic stability of soil slope by finite element limit analysis. *Computers and Geotechnics, 133*, 104074.

Li, C. D., Fu, Z. Y., Wang, Y., Tang, H. M., Yan, J. F., Gong, W. P., Yao, W. M., & Criss, R. E. (2019). Susceptibility of reservoir-induced landslides and strategies for increasing the slope stability in the Three Gorges Reservoir Area: Zigui Basin as an example. *Engineering Geology, 261*, 105279.

Li, W. J., Wang, D. W., Yang, S. F., & Yang, W. (2021b). Three Gorges Project: Benefits and challenges for shipping development in the upper Yangtze River. *International Journal of Water Resources Development, 37*(5), 758–771.

Liang, H., Yuan, W., Fu, Y., Liu, X., & Xie, Y. (2021). Degradation law of mixed-mode (I/II) fracture toughness of sandstone under drying-wetting cycles in acid and alkaline environments. *Arabian Journal of Geosciences, 14*(14), 1312.

Liu, X. R., Deng, Z. Y., Liu, Y. Q., Lu, Y. M., Liu, S. L., & Han, Y. F. (2019). Macroscopic and microscopic analysis of particle flow in pre-peak cyclic direct shear test of rock joint. *Journal of China Coal Society, 44*(7), 2103–2115. (in Chinese).

Liu, X. R., Kou, M. M., Lu, Y. M., & Liu, Y. Q. (2018a). An experimental investigation on the shear mechanism of fatigue damage in rock joints under pre-peak cyclic loading condition. *International Journal of Fatigue, 106*, 175–184.

Liu, X. R., Liu, Y. Q., Lu, Y. M., & Kou, M. M. (2020a). Experimental and numerical study on pre-peak cyclic shear mechanism of artificial rock joints. *Structural Engineering and Mechanics, 74*(3), 407–423.

Liu, Y., Dai, F., Feng, P., & Xu, N. W. (2018b). Mechanical behavior of intermittent jointed rocks under random cyclic compression with different loading parameters. *Soil Dynamics and Earthquake Engineering, 113*, 12–24.

Liu, B., Li, H. B., & Zhu, X. M. (2011). Experiment simulation study of strength degradation of rock joints under cyclic shear loading. *Chinese Journal of Rock Mechanics and Engineering, 30*(10), 2033–2039. (in Chinese).

Liu, H. L., Fei, K., & Gao, Y. F. (2003). Time history analysis method of slope seismic stability. *Rock and Soil Mechanics, 4*, 553–556+560 (in Chinese).

Liu, J., Zuo, J., Sun, Z. Y., Zillante, G., & Chen, X. M. (2013). Sustainability in hydropower development–A case study. *Renewable and Sustainable Energy Reviews, 19*, 230–237.

Liu, S. Y., Shao, L. T., & Li, H. J. (2015). Slope stability analysis using the limit equilibrium method and two finite element methods. *Computers and Geotechnics, 63*, 291–298.

Liu, X. (2010). *Research on dynamic response of slope in Wenchuan Earthquake Area*. PhD thesis. China University of Geosciences (in Chinese).

Liu, X. R., He, C. M., Liu, S. L., Liu, Y. Q., Lu, Y. M., & Liu, Z. H. (2018c). Dynamic response and failure mode of slopes with horizontal soft and hard interbeddings under frequent microseisms. *Arabian Journal for Science and Engineering, 43*(10), 5397–5411.

Liu, X. R., Jing, R., Miao, L. L., Han, Y. F., Deng, Z. Y., & Xiong, C. (2020b). Reconstruction models and typical case analysis of the fluctuation belt of reservoir bank slopes in Wushan. *Chinese Journal of Rock Mechanics and Engineering, 39*(7), 1321–1332. (in Chinese).

Liu, X. R., Liu, Y. Q., He, C. M., & Li, X. W. (2018d). Dynamic stability analysis of the bedding rock slope considering the vibration deterioration effect of the structural plane. *Bulletin of Engineering Geology and the Environment, 77*(1), 87–103.

Liu, X. R., Xu, B., Lin, G. Y., Huang, J. H., Zhou, X. H., Xie, Y. K., Wang, J. W., & Xiong, F. (2021). Experimental and numerical investigations on the macro-meso shear mechanical behaviors of artificial rock discontinuities with multiscale asperities. *Rock Mechanics and Rock Engineering, 54*, 4079–4098.

Liu, X. R., Xu, B., Liu, Y. Q., Wang, J. W., & Lin, G. Y. (2020c). Cumulative damage and stability analysis of bedding rock slope under frequent microseisms. *Chinese Journal of Geotechnical Engineering, 42*(4), 632–641. (in Chinese).

Liu, Y. Q. (2017). *Study on cumulative damage evolution mechanism and stability of bedding rock slope in reservoir area under frequent microseismic*. PhD thesis. Chongqing University (in Chinese).

Liu, Y. Q., Li, H. B., Xiao, K. Q., Li, J. C., Xia, X., & Liu, B. (2014). Seismic stability analysis of a layered rock slope. *Computers and Geotechnics, 55*, 474–481.

Lu, S. Q., & Huang, B. L. (2020). Deforming tendency prediction study on typical accumulation landslide with step-like displacements in the Three Gorges Reservoir, China. *Arabian Journal of Geosciences, 13*(9), 329.

Ma, W. T., Xu, C. P., Li, H. O., Yuan, J. L., Xu, X. W., Zhang, X. D., & Zhang, L. F. (2010). Intensive observation of reservoir-induced earthquake and preliminary analysis on the causes of earthquakes in Three Gorges Reservoir. *Seismology and Geology, 32*(4), 552–563. (in Chinese).

Mirzaghorbanali, A., Nemcik, J., & Aziz, N. (2014). Effects of shear rate on cyclic loading shear behaviors of rock joints under constant normal stiffness conditions. *Rock Mechanics and Rock Engineering, 47*(5), 1931–1938.

Nagpal, A., & Basha, B. M. (2012). Reliability analysis of anchored rock slopes against planar failure. In *Soil Proceedings of Indian Geotechnical Conference*.

Nakata, N., & Snieder, R. (2014). Monitoring a building using deconvolution interferometry. II: Ambient vibration analysis. *Bulletin of the Seismological Society of America, 104*(1), 204–213.

Ni, W. D., Tang, H. M., Liu, X., & Wu, Y. P. (2013). Dynamic stability analysis of rock slope considering vibration deterioration of structural planes under seismic loading. *Chinese Journal of Rock Mechanics and Engineering, 32*(3), 492–500. (in Chinese).

Niktabar, S. M. M., Rao, K. S., & Shrivastava, A. K. (2017). Effect of rock joint roughness on its cyclic shear behaviors. *Journal of Rock Mechanics and Geotechnical Engineering, 9*(6), 1071–1084.

Pang, R., Xu, B., Zhou, Y., & Song, L. F. (2021). Seismic time-history response and system reliability analysis of slopes considering uncertainty of multi-parameters and earthquake excitations. *Computers and Geotechnics, 136*, 104245.

Patton, F. D. (1966). Multiple models of shear failure in rock. In *Proceedings of the 1st Congress of the International Society of Rock Mechanics*.

Potyondy, D. O., & Cundall, P. A. (2004). A bonded-particle model for rock. *International Journal of Rock Mechanics and Mining Sciences, 41*(8), 1329–1364.

Qi, S. W. (2007). Evaluation of the permanent displacement of rock mass slope considering deterioration of slide surface during earthquake. *Chinese Journal of Geotechnical Engineering, 29*(3), 452–457. (in Chinese).

Ren, H. R. (2010). Spatiotemporal changes in cultivated land area and cultivated land pressure in Three Gorges Reservoir Area. *Journal of Chongqing Normal University (Natural Science), 27*(5), 23–27 (in Chinese).

Rinaldi, A. P., Improta, L., Hainzl, S., Catalli, F., Urpi, L., & Wiemer, S. (2020). Combined approach of poroelastic and earthquake nucleation applied to the reservoir-induced seismic activity in the Val d'Agri area, Italy. *Journal of Rock Mechanics and Geotechnical Engineering, 12*(4), 802–810.

Schlotfeldt, P., Elmo, D., & Panton, B. (2018). Overhanging rock slope by design: An integrated approach using rock mass strength characterisation, large-scale numerical modelling and limit equilibrium methods. *Journal of Rock Mechanics and Geotechnical Engineering, 10*(1), 72–90.

Shang, M., Xu, X., Zhang, G. D., Yi, Q. L., Ma, R., & Liu, Y. T. (2019). Back analysis of a recent progressive failure in China Three Gorges Reservoir area: Shanshucao landslide. *Journal of Testing and Evaluation, 47*(3), 1762–1776.

Siad, L. (2003). Seismic stability analysis of fractured rock slopes by yield design theory. *Soil Dynamics and Earthquake Engineering, 23*(3), 21–30.

Song, Y. X. (2019). *Dynamic evaluation of landslide risk based on integrated monitoring of satellite, airborne and ground-based data*. PhD thesis. China University of Geosciences (in Chinese).

Steedman, R. S., & Zeng, X. (1990). The influence of phase on the calculation of pseudo-static earth pressure on a retaining wall. *Geotechnique, 40*(1), 103–112.

Stianson, J. R., Chan, D., & Fredlund, D. G. (2015). Role of admissibility criteria in limit equilibrium slope stability methods based on finite element stresses. *Computers and Geotechnics, 66*, 113–125.

Sun, G. H., Lin, S., Zheng, H., Tan, Y. Z., & Sui, T. (2020). The virtual element method strength reduction technique for the stability analysis of stony soil slopes. *Computers and Geotechnics, 119*, 103349.

Tang, H. M., Wasowski, J., & Juang, C. H. (2019). Geohazards in the three Gorges Reservoir Area, China-Lessons learned from decades of research. *Engineering Geology, 264*, 105267.

Tao, Z. G., Geng, Q., Zhu, C., He, M. C., Cai, H., Pang, S. H., & Meng, X. Z. (2019). The mechanical mechanism of large-scale toppling failure for counter-inclined rock slopes. *Journal of Geophysics and Engineering, 16*(3), 541–558.

Tao, Z. G., Zhu, C., He, M. C., & Karakus, M. (2021). A physical modeling-based study on the control mechanisms of negative Poisson's ratio anchor cable on the stratified toppling deformation of anti-inclined slopes. *International Journal of Rock Mechanics and Mining Sciences, 138*, 104632.

Tu, Y. L., Liu, X. R., Zhong, Z. L., & Li, Y. Y. (2016). New criteria for defining slope failure using the strength reduction method. *Engineering Geology, 212*, 63–71.

Wan, L. P., Zhou, M. L., & Desar, S. (2017). Long-term stability calculation of reservoir bank slope considering water–rock interaction. *Tehnički Vjesnik, 24*(1), 283–289.

Wang, C. X. J., Deng, H. F., Zhang, H. B., Fang, J. C., & Xiao, Y. (2018). Seismic behavior of a typical bank slope of Three Gorges Reservoir in consideration of degradation of rock mass mechanical properties. *Journal of Changjiang River Scientific Research Institute, 35*(1), 112–116. (in Chinese).

Wang, G. L., Zhang, C. H., Peng, G., & Chen, C. W. (1994). Dynamic test of rigid blocks and parameter study in dynamic analysis of distinct elements. *Chinese Journal of Rock Mechanics and Engineering, 13*(2), 124–133. (in Chinese).

Wang, H. B., Sun, Y. H., Tan, Y. Z., Sui, T., & Sun, G. H. (2019). Deformation characteristics and stability evolution behavior of Woshaxi landslide during the initial impoundment period of the Three Gorges reservoir. *Environment and Earth Science, 78*(20), 592.

Wang, L. Q., Yin, Y. P., Zhou, C. Y., Huang, B. L., & Wang, W. P. (2020). Damage evolution of hydraulically coupled Jianchuandong dangerous rock mass. *Landslides, 17*(5), 1083–1090.

Wang, R. S. (2017) Induced earthquake monitoring and forecast of Three Gorges Reservoir. *Journal of China Three Gorges University (Natural Sciences), 39*(1), 30–37 (in Chinese).

Wang, S. J., & Zhang, J. M. (1982). On the dynamic stability of block sliding on rock slopes. *Scientia Geologica Sinica, 2*, 162–170. (in Chinese).

Wang, Y. X. (2006). *Study on the effect of water–rock interaction on stability of reservoir bank.* Master's thesis. Chongqing University (in Chinese).

Wang, Z. J., Liu, X. R., Yang, X., & Fu, Y. (2017). An improved Duncan-Chang constitutive model for sandstone subjected to drying-wetting cycles and secondary development of the model in FLAC (3D). *Arabian Journal for Science and Engineering, 42*(3), 1265–1282.

Xia, Y., Fang, Z. F., Zou, F., Yu, C. H., & Zhao, L. H. (2021). Pseudo-dynamic analysis of the influence of structural plane deterioration characteristics on the stability of rock slope. *Highway, 66*(4), 67–71. (in Chinese).

Xiong, F., Liu, X. R., Zhou, X. H., Lin, G. Y., Liu, D. S., Han, Y. F., Xu, B., He, C. M., & Wang, Z. J. (2022). Mechanical behaviours of sandstone containing intersecting fissures under uniaxial compression. *Journal of Rock Mechanics and Geotechnical Engineering, 14*(2), 460–476.

Xu, B., Liu, X. R., Zhou, X. H., Guo, X. Y., Wang, Y., Huang, J. H., Liu, J., Xiong, F., & Zhang, J. L. (2022a). Investigation on the dynamic cumulative damage mechanism and stability of bedding rock slope under the deterioration of rock mass in the hydro-fluctuation belt. *Bulletin of Engineering Geology and the Environment, 81*(8), 332.

Xu, B., Liu, X. R., Zhou, X. H., Huang, J. H., Wang, L. F., Lin, G. Y., Wang, J. W., & Zhang, J. L. (2022b). Investigations on the macro-meso cumulative damage mechanism of the discontinuities in soft-hard interbedded rock mass under pre-peak cyclic shear loading. *International Journal of Rock Mechanics and Mining Sciences, 158*, 105184.

Xu, B., Liu, X. R., Zhou, X. H., Liu, J., Huang, J. H., Wang, Y., & Zeng, X. (2022c). Experimental study on dynamic response law of bedding rock slopes under deterioration of rock mass in hydro-fluctuation belt. *Chinese Journal of Geotechnical Engineering, 44*(8), 1453–1462. (in Chinese).

Xu, B., Liu, X. R., Zhou, X. H., Xie, Y. K., Suliman, L., Liu, X. L., Lin, G. Y., & Huang, J. H. (2021). Investigation on the macro-meso fatigue damage mechanism of rock joints with multiscale asperities under pre-peak cyclic shear loading. *Soil Dynamics and Earthquake Engineering, 151*, 106958.

Xu, G. L., Li, W. N., Yu, Z., Ma, X. H., & Yu, Z. Z. (2015). The 2 September 2014 Shanshucao landslides, Three Gorges Reservoir, China. *Landslides, 12*(6), 1169–1178.

Yang, C. W., Guo, X. Y., Liu, Y., & Wang, D. (2021). Research on dynamic response and instability failure phenomenon of deposit slope with weak interlayer based on shaking table test. *Advanced Engineering Science, 53*(3), 43–52. (in Chinese).

Yang, Q., Wang, Y. L., & Ma, Y. Y. (2020a). Distribution rule and influencing factors of geological disasters from 2001 to 2019 in China. *Journal of Geological Hazards and Environment Preservation, 31*(4), 43–48. (in Chinese).

Yang, Z. P., Lai, Y. L., Liu, S. L., Tian, X., Hu, Y. X., & Ren, S. P. (2019). Dynamic stability and failure mode of slopes with overlying weak rock mass under frequent micro-seismic actions. *Chinese Journal of Geotechnical Engineering, 41*(12), 2297–2306. (in Chinese).

Yang, Z. P., Tian, X., Jiang, Y. W., Liu, X. R., Hu, Y. X., & Lai, Y. L. (2020b). Experimental study on dynamic characteristics and dynamic responses of accumulation slopes under frequent microseisms. *Arabian Journal of Geosciences, 13*(16), 770.

Yao, X. F., Hu, X., Yi, L. X., Zeng, X. X., & Zhang, L. N. (2017a). Estimation of Coulomb stress changes and induced earthquake occurrences prediction in the region of Three Gorges Reservoir, China. *Environmental Earth Sciences, 76*(24), 826.

Yao, Y. S., Wang, Q. L., Liao, W. L., Zhang, L. F., Chen, J. H., Li, J. G., Yuan, L., & Zhao, Y. N. (2017b). Influences of the Three Gorges Project on seismic activities in the reservoir area. *Science Bulletin, 62*(15), 1089–1098.

Ye, R. Q., Fu, X. L., Guo, F., Yi, Q. L., Zhang, J. Y., Li, C. M., Hou, S. P., & Liu, N. (2021). Deformation characteristics and mechanism analysis of geological hazards during operation period of Three Gorges Reservoir. *Journal of Engineering Geology, 29*(3), 680–692 (in Chinese).

Ye, S. H., & Huang, A. P. (2020). Sensitivity analysis of factors affecting stability of cut and fill multistage slope based on improved grey incidence model. *Soil Mechanics and Foundation Engineering, 57*, 8–17.

Yin, Y. P., Huang, B. L., Wang, W. P., Wei, Y. J., Ma, X. H., Ma, F., & Zhao, C. J. (2016). Reservoir-induced landslides and risk control in Three Gorges Project on Yangtze River, China. *Journal of Rock Mechanics and Geotechnical Engineering, 8*(5), 577–595.

Yin, Y. P., Huang, B. L., Zhang, Q., Yan, G. Q., & Dai, Z. W. (2020). Research on recently occurred reservoir-induced Kamenziwan rockslide in Three Gorges Reservoir, China. *Landslides, 17*(8), 1935–1949.

Yin, Y. P., Wang, H. D., Gao, Y. L., & Li, X. C. (2010). Real-time monitoring and early warning of landslides at relocated Wushan Town, the Three Gorges Reservoir, China. *Landslides, 7*(3), 339–349.

Yoon, J. (2007). Application of experimental design and optimization to PFC model calibration in uniaxial compression simulation. *International Journal of Rock Mechanics and Mining Sciences, 44*(6), 871–889.

Zang, M. D., Yang, G. X., Dong, J. Y., Qi, S. W., He, J. X., & Liang, N. (2022). Experimental study on seismic response and progressive failure characteristics of bedding rock slopes. *Journal of Rock Mechanics and Geotechnical Engineering, 14*(5), 1394–1405.

Zhang, Y. C., Jiang, Y. J., Asahina, D., & Wang, C. S. (2020). Experimental and numerical investigation on shear failure behaviors of rock-like samples containing multiple non-persistent joints. *Rock Mechanics and Rock Engineering, 53*(10), 4717–4744.

Zhang, B., & Chen, H. Q. (2001). Analysis of abutment aseismatic stability by using finite element and rigid limit equilibrium method. *Chinese Journal of Rock Mechanics and Engineering, 20*(5), 665–670. (in Chinese).

Zhang, C. Y., Yin, Y. P., Yan, H., Li, H. X., Dai, Z. W., & Zhang, N. (2021). Reactivation characteristics and hydrological inducing factors of a massive ancient landslide in the three Gorges Reservoir, China. *Engineering Geology, 292*, 106273.

Zhang, F., Li, J. G., Wei, G. C., Liao, W. L., Wang, Q. L., & Xiang, C. F. (2017). Analysis of the relationship between water level fluctuation and seismicity in the Three Gorges Reservoir (China). *Geodesy and Geodynamics, 8*(2), 96–102.

Zhang, Y. C., Deng, H. F., Wang, W., Duan, L. L., Zhi, Y. Y., & Li, J. L. (2018). The dynamic response law of bank slope under water–rock interaction. *Advances in Civil Engineering, 2018*, 1306575.

Zhao, P. (2009). *Study on the reservoir induced seismicity in Chongqing under the running of the Three Gorges Reservoir*. Master's thesis. Chongqing Jiaotong University (in Chinese).

Zheng, Y. R., Ye, H. L., Huang, R. Q., Li, A. H., & Xu, J. B. (2010). Study on the seismic stability analysis of a slope. *Journal of Earthquake Engineering and Engineering Vibration, 30*(2), 173–180. (in Chinese).

Zhu, C., He, M. C., Karakus, M., Zhang, X. H., & Tao, Z. G. (2021). Numerical simulations of the failure process of anaclinal slope physical model and control mechanism of negative Poisson's ratio cable. *Bulletin of Engineering Geology and the Environment, 80*(4), 3365–3380.

Zhu, C. H. (2010). *Study on data warehouse and data mining application on geohazard in Three Gorges Reservoir Area*. PhD thesis. China University of Geosciences (in Chinese).

MIX
Papier aus verantwortungsvollen Quellen
Paper from responsible sources
FSC® C105338

If you have any concerns about our products,
you can contact us on
ProductSafety@springernature.com

In case Publisher is established outside the EU,
the EU authorized representative is:
Springer Nature Customer Service Center GmbH
Europaplatz 3, 69115 Heidelberg, Germany

Printed by Libri Plureos GmbH
in Hamburg, Germany